Issues and Cases of Degrowth in Tourism

Issues and Cases of Degrowth in Tourism

Edited by

Konstantinos Andriotis
Professor in Tourism
Department of Marketing, Branding and Tourism
Middlesex University
UK

CABI

CABI is a trading name of CAB International

CABI
Nosworthy Way
Wallingford
Oxfordshire OX10 8DE
UK

Tel: +44 (0)1491 832111
Fax: +44 (0)1491 833508
E-mail: info@cabi.org
Website: www.cabi.org

CABI
WeWork
One Lincoln St
24th Floor
Boston, MA 02111
USA

Tel: +1 (617) 682 9015
E-mail: cabi-nao@cabi.org

A catalogue record for this book is available from the British Library, London, UK.

Library of Congress Cataloging-in-Publication Data

Names: Andriotis, Konstantinos, editor.
Title: Issues and cases of degrowth in tourism / editor, Konstantinos Andriotis.
Description: Wallingford, Oxfordshire ; Boston, MA : CABI, 2021. | Includes bibliographical references and index. | Summary: "Degrowth in tourism is the voluntary shift to rebuild destinations and local economies where consumption, production and the exploitation of resources are minimal. This book takes steps to address the paucity of combined research on tourism and degrowth by presenting emergent knowledge and research on the dynamics of degrowth from a tourism perspective by presenting cases from different parts of the world"-- Provided by publisher.
Identifiers: LCCN 2020036453 (print) | LCCN 2020036454 (ebook) | ISBN 9781789245073 (hardback) | ISBN 9781789245080 (ebook) | ISBN 9781789245097 (epub)
Subjects: LCSH: Sustainable tourism. | Overtourism.
Classification: LCC G156.5.S87 I87 2021 (print) | LCC G156.5.S87 (ebook) | DDC 338.4/791--dc23
LC record available at https://lccn.loc.gov/2020036453
LC ebook record available at https://lccn.loc.gov/2020036454

ISBN-13:	9781789245073	(hardback)
	9781789245080	(ePDF)
	9781789245097	(ePub)

Commissioning Editor:	Claire Parfitt
Editorial Assistant:	Lauren Davies
Production Editor:	Tim Kapp

Typeset by Exeter Premedia Services Pvt Ltd, Chennai, India
Printed and bound in the UK by Severn, Gloucester

Contents

Contributors

Konstantinos Andriotis, Middlesex University, London, UK. E-mail: k.andriotis@mdx.ac.uk

Ernest Cañada, Alba Sud - Investigation and Communication for Development, Barcelona, Spain. E-mail: ernest@albasud.org

Brendan Canavan, University of Huddersfield, West Yorkshire, UK. E-mail: b.canavan@hud.ac.uk

Martin Fontanari, ISM International School of Management GmbH, Cologne, Germany. E-mail: martin.fontanari@ism.de

Neil Hughes, University of Nottingham, Nottingham, UK. E-mail: neil.hughes@nottingham.ac.uk

José Mansilla, University of Lleida, Barcelona, Spain. E-mail: joseamansilla@hotmail.com

Pascal Mayer, Université de Pau et des Pays de l'Adour, France. E-mail: pas.mayer@laposte.net

Sabine Panzer-Krause, University of Hildesheim, Hildesheim, Germany. E-mail: sabine.panzer-krause@uni-hildesheim.de

Dalia Perkumienė, Vytautas Magnus University Agriculture Academy, Lithuania. E-mail: dalia.perkumiene@vdu.lt

Ammalia Podlaszewska, Bauhaus-Universität Weimar, Weimar, Germany. E-mail: ammalia.podlaszewska@gmail.com

Rasa Pranskūnienė, Vytautas Magnus University Agriculture Academy, Lithuania. E-mail: rasa.pranskuniene@vdu.lt

Hugues Seraphin, University of Winchester, Winchester, UK. E-mail: hugues.seraphin@winchester.ac.uk

Anastasia Traskevich, Belarusian State Economic University, Belarus. E-mail: anastasia.traskevich@akalem.de

1 Introduction

Konstantinos Andriotis*

Middlesex University, London, UK

1.1 Introduction

With 1.4 billion international tourist arrivals and an $8.8 trillion contribution to the global economy in 2018 (UNWTO, 2019), as well as forecasts which predict an increase of tourists' movements by 2029 to almost 2.2 billion (an increase of 3.8% pa) and capital investment to rise from $940.9 billion in 2018 to $1,489.5 billion in 2029 (4.2% pa), travel and tourism is one of the world's largest sectors and one of the major cultural and economic forces on the planet. Nevertheless, while past experiences indicate that the tourism industry is addicted to growth, forecasts have been recently revised downward by unforeseen disasters and crises attributed to the COVID-19 (coronavirus) outbreak that was an unexpected shock for the global tourism industry, especially as the fear factor restricted tourists' mobility and governments closed their borders in their effort to curtail and control the pandemic (J.P. Morgan Chase and Co, 2020; Andriotis and Paraskevaides, forthcoming).

Prior to the COVID-19 pandemic, the well-established global tourism system had for several years grappled with concerns of overtourism, but as a result of the virus has within a short time moved from overtourism to nontourism (Gössling *et al.*, 2020; Higgins-Desbiolles, 2020). In popular destinations where overcrowding was widely recognized, a complete societal lockdown to combat the spread of COVID-19 plummeted tourism worldwide, leaving millions of tourism employees and businesses economically paralysed. Such conditions rapidly decreased mobile world populations and while this has obviously reduced carbon emissions, at the same time it has resulted in the collapse of the whole tourism sector, with airlines cancelling their flights and countries closing their borders.

However, it is not only viruses that should make us more concerned travellers and citizens. In many countries and destinations, tourism has grown beyond sustainable bounds. Endless tourism growth and various catastrophic human activities need to change to avoid the worst effects of

*k.andriotis@mdx.ac.uk

DOI: 10.1079/9781789245073.0001

human-induced global issues. Tourism activities, along with several other human pursuits, have caused the rise in carbon emissions which has resulted in global warming and climate change, both of which have caused intense and unpredictable weather events affecting several destinations across the globe. The negative effects of increased tourist movements have been felt by host communities, the tourists themselves and the environment and this, in conjunction with the failure of destination decision makers to manage tourism in a sustainable way, has directed several scholars (e.g. Brenner and Theodore, 2002; Jonas and Gibbs, 2003; Andriotis, 2003a; Andriotis and Vaughan, 2009; Campbell, 2016) to criticize the term 'sustainable tourism' for failing to alleviate problems facing developed and developing countries as a result of unplanned and/or unsustainable expansion of their tourism industry. In practice, by supporting hegemonic, capitalist relationships and serving as a 'flanking mechanism' to neoliberalism, the tourism industry has been incompatible with sustainability goals (Higgins-Desbiolles, 2018, p. 157).

In this respect, degrowth through a controlled decrease in economic activity came to the foreground as a way to meet host communities' needs by ensuring a fair distribution of wealth and environmental use of resources (Latouche, 2006). Degrowth as a way of life has been adopted by several primitive societies through the centuries and can be traced back to ancient Greece and Rome, where philosophers, such as Diogenes the Cynic, voluntarily embraced a meaningful life by returning to the simplicity of nature. Over recent centuries, various degrowth debates have been ongoing, expressed first by critical thinkers such as Marx (1859) and more recently by Georgescu-Roegen (1971, 1979), Illich (1973) and Schumacher (1973), who challenged the conventional economic concept of unlimited growth. Despite its long history the term 'degrowth' emerged only after 2008, building from critical scholarship as well as activist social movements (Demaria *et al.*, 2011, p. 2).

From a tourism perspective, degrowth in line with the post-tourists or postmodern tourists who embrace openly, 'but with some irony, the increasingly inauthentic, commercialised and simulated experiences offered by the tourism industry' (Smith *et al.*, 2010, p. 129), has been explored as an ideology of opposition to conventional mass tourism and the exploitation of local communities and their environmental and cultural resources. As a philosophical concept, it has been expressed from the counterculture of youths and younger hippies during the mid-1960s, who chose to turn their back on the system and values of industrial society by visiting pristine spaces open to conditions of freedom distinctly different from the normal place of home (Andriotis, 2006, 2013), as well as several forms of degrowth-inspired travellers such as 'drifters' (Cohen, 1973), 'wanderers' (Vogt, 1976), 'neo nomads' (MacCannell, 1992), 'tribal tourists' (Gibson *et al.*, 2013), 'new age travelers' (Martin, 1998, 2000), 'frontier travelers' (Laing, 2006; Laing and Crouch, 2009a, b) and antinomians (Andriotis, 2013). These degrowth-inspired forms of travelling have emerged from people who choose destinations that are not affected by commercialization and overconsumption, but instead their personal time is focused on the richness of the experience through travelling less, more slowly and using low carbon options; all of which aim to protect the environment and the local culture, as well as support the local economy.

Despite the fact that several types of tourist, such as those mentioned above, have followed several principles of degrowth, as well as the admittedly limited attempts of some host communities and local authorities to 'rightsize' their tourism industry, there is a lack of research identifying and critically assessing the discourse of tourism degrowth itself. This paucity of research combining degrowth and tourism has hindered our understanding about degrowth-induced tourism development and travel. It is the aim of this book to scientifically address this paucity by presenting case studies on the dynamics of degrowth from different parts of the world.

1.2 COVID-19 Recession vs Degrowth

As a result of the COVID-19 crisis and the consequent restrictions on mobility, the tourism industry ground to a halt (Fletcher *et al.*, 2020). Nations shut their borders; flights, festivals and events were cancelled; airports, ports, resorts, hotels and food and beverage outlets closed; tourism and associated businesses faced bankruptcy; many employees in the tourism sector and all sectors dependent on tourism suffered from loss of income or even lost their jobs and, due to fear and uncertainty, many people changed their travel choices by either avoiding non-essential travel or choosing destinations in the vicinity of their home (Andriotis and Paraskevaides, forthcoming). As WTTC (2020) reports, due to the COVID-19 pandemic and the associated travel restrictions, the economy of many host communities worldwide has been threatened. In more detail, UNWTO (2020) estimated that the COVID-19 pandemic would cause international tourist arrivals to decline by 60–80% (compared with 2019 figures) rather than the forecasted 3–4% growth. This decline is expected to result in a loss of US$910 billion to US$1.2 trillion in export revenues and 100 to 120 million direct tourism jobs (UNWTO, 2020).

While globalization made possible people's free movement, at the same time it predisposed the COVID-19 pandemic (Nelson and Liegey, 2020). Several pandemics, including COVID-19, have spread rapidly due to tourists' movements and the use of planes and cruise ships. In the words of Monterrubio (2010):

> international travel has been regarded as an important factor contributing to emerging infections. This is because international travel and commerce have modified the size and mobility of human populations, bringing some environments, humans and other animal species into contact with each other like never before (Tapper, 2006) ... Owing to the great development and increase in air transportation, the international and in-flight spread of contagious diseases by air travellers have been reported as ways in which infections can be transmitted (Mangili and Gendreau, 2005) (pp. 2–3).

Hyper-mobile capitalist societies encouraged by a) frequent long-haul travel of westerners to exotic resorts for a week's holiday; b) short-distance flights, for example from Italy to Austria for weekend ski vacations and c) massive cruise ship movements, have exacerbated the spread of COVID-19. For instance, the first infected diagnosed incidents of COVID-19 in Greece were religious travellers returning from Israel and one of the major outbreaks took place on board the *Diamond Princess* cruise ship when over 700 people became infected, and 12 people died, and therefore the ship was quarantined with 3711 passengers and crew members on board (The *Guardian*, 2020).

While the reasons for 'slowing down' that degrowth advocates (mainly reducing global emissions) differ from COVID-19 imperatives (i.e. a reduction of the spread of infection), both lead us to similar conclusions: fly less and mindfully, move more slowly and spend much more time in our communities (including staycations). All these are actually needed not only to restrict the pandemic but also for a number of ecological and social reasons, as various degrowth advocates assert (e.g. Latouche, 2010; Schneider *et al.*, 2010; Kallis, 2011; Alexander, 2012, 2014; Andriotis, 2014, 2018). Thus the current economic contraction of the COVID-19 pandemic can be used as a starting point for transforming the global travel system in a direction that could better serve the needs of the planet.

While pollution and emissions have reduced following COVID-19 lockdowns, the unplanned downscaling of social and economic activity cannot be associated with degrowth. As Alexander (2020) highlights, degrowth means *planned* economic contraction, but nothing about the existing lockdowns in response to COVID-19 worldwide was planned. The involuntary and unplanned economic contraction that has had such negative socioeconomic effects, such as increased unemployment and poverty, should not be confused with degrowth but regarded as a part of recession or, if it persists, depression (Alexander, 2020). In practice, degrowth requires intentional and democratic transformation. Therefore, the recession resulting from COVID-19 should not be confused with degrowth (Fletcher *et al.*, 2020).

1.3 Overtourism and Growth vs Limits to Growth and Degrowth

Following residents' suffering due to overtourism, an atmosphere of *tourismphobia* has been generated in several European cities, such as Venice, Barcelona, Dubrovnik, Amsterdam, Reykjavik and Santorini, resulting in anti-tourism movements (Hall, 2011; Coldwell, 2017; Boluk *et al.*, 2019; Milano *et al.*, 2019a; Panzer-Krause, 2019; Valdivielso and Moranta, 2019). Residents want to reclaim their lifestyle by refusing to interact with a high volume of visitors and discouraging tourists from visiting their communities. To achieve this, they call for a reduction in the number of visitors and stricter regulation of their tourism industry, as well as a slowdown in the growth of tourism (Seraphin *et al.*, 2020). Just to name one example, Venice, in an attempt to reduce visitor numbers, introduced in 2018 entrance fees to the city (Brown, 2019), although considering the negative economic effects of COVID-19 on the Italian economy continued enforcement of this measure might need to be postponed.

Concerns with growth have a long history. Even in 300 BC the Greek philosopher Plato complained about landscape changes in Attica (Janssen *et al.*, 1995; Andriotis, 2000). However, it was only after the 1970s, with the Club of Rome's limits to growth report (Meadows *et al.*, 1972), that neoliberal capitalism, which is reliant on unlimited growth, was disputed. Following this report and in an attempt to overcome excessive visitation to tourist destinations, radical proposals such as that of degrowth found increasing attention, leading scholars such as Andriotis (2018) and Valdivielso and Moranta (2019) to call for a shift from overtourism to degrowth.

Led by the environmental movement, degrowth requires key strategies aiming to balance tourism growth by reducing consumption and production and ensuring institutional changes and technological development in harmony with the environment (Hinrichsen, 1987; BonJour, 2009). These degrowth measures came to the foreground for three reasons. First, from an economic perspective, with millions of people unemployed after the economic crisis that started in 2007, several orthodox economists, such as Larry Summers and Paul Krugman, realized that while in the past the only solution to poverty and several other problems faced by economies worldwide was more growth, today they have to face the 'end of growth' (Bonaiuti, 2014), since overproduction and overconsumption is putting our planet and society at risk. While economic recession is not equivalent to degrowth, since it involves involuntary or forced reduction of material consumption (Buchs and Koch, 2019), it directed several scholars to think about the necessity of degrowth through reconsideration of existing models of economic growth and a need to adopt measures toward desirable growth.

From an environmental perspective, despite arguments made in the past by tourism advocates portraying tourism as a relatively 'clean' and 'benign' industry whose growth is associated with increased economic welfare and job creation (Andriotis and Vaughan, 2004; Andriotis, 2005, 2008; Stroebel, 2015), it has been proved that its adverse environmental effects are considerable to the extent that Valdivielso and Moranta (2019) have denounced tourism as an agent of territorial urbanization, attributed to overtourism and the associated large number of tourists visiting the same place at the same time (Seraphin *et al.*, 2020). This excessive visitation is usually encouraged by the rich and unique attractions of a destination and its sophisticated marketing initiatives, and fuelled by low-cost flights and the popularization of home-sharing platforms, such as Airbnb, which offer cheaper accommodation options compared to hotels (Andriotis, 2018, p. 12).

From a sociocultural perspective the fact that an increasing number of tourists desire to see the 'real' and 'authentic' everyday life of the host population, not only in the main tourist areas of urban destinations, but also outside (Andriotis, 2003b, 2009, 2011; Pappalepore *et al.*, 2014), reinforces pressures on the sociocultural resources of host communities. As a result, several cases have been reported where reactions have taken place against incoming tourists and tourism expansion by community members who express open resistance in the form of social movements, anti-tourist protest and agitation (Andriotis, 2018) – as Hughes and Mansilla's contribution in this book reports in the case of Barcelona (see also Kousis, 2000; Navarro-Jurado *et al.*, 2019).

As Fletcher *et al.* (2019) remark, degrowth is *not* against tourism itself. It requires the tourism industry to be differently scaled and organized based on the fact that it is not possible to have limitless economic growth on a finite planet (Andriotis, 2002; Tukker, 2014). In practice, 'the formalization of degrowth as a new economic paradigm aiming to reduce entropy that results from mass production and consumption' (Andriotis, 2018, p. 14) should be directed to the voluntary downsizing of the tourism industry, the final outcome of which will be the rightsizing and steady state of the industry (Hall, 2009, 2010) and a non-growth society with a stable throughput of resources (Daly, 1991). This is the main reason that degrowth differs from sustainable development as the section below explains.

1.4 Degrowth vs Sustainability

A plethora of strands (e.g. neoliberalism, dependency, diffusion, modernization and sustainability) have influenced and continue to influence development theory in the case of tourism (Andriotis, 2003a; Andriotis and Vaughan, 2009). Among them the concept of sustainable development has shaped in the last decades the way several scholars think about the relationship between economic growth and environmental degradation (Johnsen *et al.*, 2017, p. 4), while the newly developed paradigm of degrowth, in spite of its undoubted significance for destination development, has received a weak application in the study of tourism. In fact, although scholars have paid tremendous attention over the last few years to overtourism they have neglected the significance of degrowth as a strategy to address problems accrued from overtourism and overdevelopment as well as not anticipating or trying to prevent any potential problems from destinations being developed. Thus, despite the recent research interest in degrowth from tourism scholars, the concept still remains open to multiple interpretations.

At first glance, degrowth has much in common with the most widely accepted definition of sustainable development provided by the World Commission on Environment and Development (WCED) in the Brundtland Report, *Our Common Future* (WCED, 1987), since both propose the preservation of resources and emphasize the need to respect future generations as a basis for development (Andriotis, 2018). Nevertheless, attempts made to subsume degrowth under the rubric of sustainable development have received severe criticism from degrowth scholars (Latouche, 2009, p. 9). As Fletcher *et al.* (2019) point out: 'despite decades of concerted global effort to achieve sustainable development, socioecological conflicts and inequality have rarely reversed, but in fact increased in many places' (p. 14). Apparently, sustainable development has not managed to deliver the promised results. For this reason, Reichel (2015) proposes the replacement of the concept of sustainable development with sustainable degrowth under which ecological and social-minded activists and researchers might rally. This term, which has evolved into a prominent theme in the last decade, is defined as 'a democratic, equitable and environmentally beneficial process of gradually decreasing the volume of production and consumption, the goal of which is to improve human wellbeing' (Fraňková and Johanisová, 2013, p. 13).

The need to replace the term sustainable development with sustainable degrowth emerged because sustainable development has not succeeded in truly delivering its promises to reconcile social equity and respect the limitations of planetary ecosystems (Reichel, 2015) through a decrease in the volume of production and consumption. This is because sustainability is not *against* the quest for continuous growth but, instead, it contends that only with economic growth, and under increasing productive potential, can the aims of solving environmental and social threats be worked out (Owen *et al.*, 1993; Janssen *et al.*, 1995; Andriotis, 2018). In the words of Chassagne and Everingham (2019): 'sustainable tourism is concerned with the limits to growth, but it does not go far enough in reconsidering those limits. While sustainable tourism argues for a shift towards social and environmental responsibility, it is still couched within the unsustainable economic growth paradigm that inevitably has negative effects on the environment and local communities' (p. 24). With this in mind, Banerjee (2008) accuses sustainable development of serving more as a slogan rather than as a concept, that has gradually replaced the slogan of economic growth. In fact, while both concepts (sustainability and degrowth) may share a common root, degrowth proposes the end of economic growth as we know it (Klitgaard and Krall, 2012), though sustainability suggests the end of unbalanced growth (Perles-Ribes *et al.*, 2018).

Thus, degrowth blames the growth orientation of sustainability on the basis that it is not always ecologically and culturally sustainable, and may eventually destroy the resources of destinations. In fact, sustainability does not challenge unbridled consumerism, but instead places emphasis on technology and better management etc. as ways to balance tourism development. This, in conjunction with the fact that cases of strong sustainable development as mentioned by Hunter (1997) are very limited, means that 'true sustainable tourism development' is unachievable (Sharpley, 2002). Thus, degrowth comes as a popular response to the failure of sustainable development by paying attention to the issues of limits to growth and of equal distribution of our shared inheritance in a move from a growth-based economy to a voluntary shift that will rebuild local economies in a way where consumption, production and the exploitation of resources is minimal.

1.5 Past Research on Degrowth in Tourism

A few years ago, the term 'degrowth' was hardly used by tourism scholars. While the term had attracted a great deal of research attention from various disciplines, such as environmental studies, political and social sciences, economics, history, anthropology and technology, only in the last two years have tourism publications on this topic increased, indicating tourism scholars' interest in exploring the concept of degrowth from a tourism perspective. Despite these attempts, degrowth has not yet efficiently been explored in the case of the tourism industry.

A Scopus database search in March 2020 for the keywords 'tourism' and 'degrowth' in titles, abstracts and keywords identified a total of 23 studies. (This number is small compared to the 274 studies listed in response to the term 'overtourism', a term that has been widely used since 2016 – Capocchi *et al.*, 2020.) Exclusion of irrelevant publications and cleaning of raw data sets resulted in a total of ten studies having an explicitly tourism focus. Also, four studies were added through the authors' extensive research on the topic over the last ten years. (Table 1.1 lists all degrowth studies that have a tourism focus and presents details about their methodological approach, aims and key findings.)

From the 14 published studies only five had been published prior to 2018 and the rest after 2019. The majority (total eight papers) came from a special issue edited by Fletcher, Blázquez-Salom, Murray and Blanco-Romero of the *Journal of Sustainable Tourism*. With the exception of one conference paper included in proceedings (Andriotis, 2014) and one monograph (Andriotis, 2018), all other studies were published in journals. From a review of these studies it was evident that five were theoretical/conceptual and nine had collected primary data mainly from host communities. (The exception was the study of Andriotis (2013) who conducted in-depth interviews with visitors to a degrowth destination.) In line with Weiss and Cattaneo's (2017) review of peer-refereed degrowth literature where many studies used Spain as a case study, the review of tourism studies indicates that research was also dominated by Spanish cases (four in total). There were also two papers for Ecuador. The majority of articles (eight in total) were written by a single author. All studies that collected primary data had adopted a qualitative approach. This was expected since the topic is under-researched and therefore the studies were exploratory and aimed to gain further insights on various topics associated with degrowth in tourism.

Based on the review of past research, the contribution of this book is threefold. First, it presents cases from countries such as Indonesia and Germany, where published research on the topic is limited or non-existent in the English language. Second, it overcomes past research negligence by covering under-researched aspects of tourism degrowth such as the role of tour operators, corporate social responsibility (CSR), heterotopia, freedom of movement, etc. Third, following Fletcher *et al.*'s special issue of the *Journal of Sustainable Tourism* and the book by Andriotis (2018), this edited volume is among limited attempts to initiate and stimulate degrowth in the global debate of tourism academic research.

1.6 Structure of the Book

The book aims to scientifically address the paucity of combined research on tourism and degrowth by presenting case studies on the dynamics of

Table 1.1. Past research on degrowth in tourism.

	Year	Title	Author(s)	Methodological approach	Aim	Key finding(s)
1.	2019	Networking towards sustainable tourism: innovations between green growth and degrowth strategies	Panzer-Krause (2019)	• Network analysis (semi-structured interviews with stakeholder groups) • Case: Burren and Cliffs of Moher region, Ireland	Investigates features of a regional network of tourism enterprises in rural Ireland	Degrowth strategists need to be represented as key actors of communication and joint activities in order to realize a clear shift away from business as usual
2.		The social construction of the tourism degrowth discourse in the Balearic Islands	Valdivielso and Moranta (2019)	• Critical discourse analysis (participant observation and secondary sources) • Case: Balearic Islands, Spain	Analyses the debate around tourism degrowth in the Balearic Islands between 2014 and 2019	The debate over tourism degrowth is not just a discrepancy between different approaches dealing with overtourism, tourism pressure and gentrification, but a social struggle where new democratic political subjects can be coalesced and organized in a civil society
3.		Tourism and degrowth: an emerging agenda for research and praxis	Fletcher *et al.* (2019)	Conceptual/theoretical	Explores the potential of degrowth to facilitate a truly sustainable tourism	Tourism degrowth is a proposal that brings social movements and research/praxis together. Scholars need to pay attention to and cooperate with social movements to make tourism comply with planetary boundaries

Continued

Table 1.1. Continued

	Year	Title	Author(s)	Methodological approach	Aim	Key finding(s)
4.		Territorial tourism planning in Spain: from boosterism to tourism degrowth?	Blázquez-Salom *et al*. (2019)	• Qualitative/grounded theory (secondary sources, participant observation, interviews with stakeholder groups) • Case: mature coastal destinations in Spain	Examines the progressive incorporation of various measures in territorial tourism planning	In Spain, the traditional emphasis on urban tourism growth is declining and more restrictive policies are now being implemented. There is a need for much stronger measures to pursue genuine and fair degrowth without systemic constraints
5.		Tourism as a right: a frivolous claim against degrowth?	Gascón (2019)	Analysis of primary documentation	Analyses an action promoted by the UNWTO to turn tourism into a human right	By legitimizing tourism as a supposed human right, it would allow the debate to centre on a conflict of rights
6.		Overtourism and degrowth: a social movements perspective	Milano *et al*. (2019b)	• Ethnographic research (rapid situation analysis – participatory method based on in-depth semi-structured interviews, participant observation, informal conversations and retrospective evaluation of field diary entries) • Case: Barcelona, Spain	Explores the evolution of the tourism degrowth discourse among social movement activists in Barcelona	To achieve a substantive paradigm shift in urban tourism governance, both technical and political solutions against overtourism are required

Continued

Table 1.1. Continued

	Year	Title	Author(s)	Methodological approach	Aim	Key finding(s)
7.		Buen Vivir (BV): Degrowing extractivism and growing wellbeing through tourism	Chassagne and Everingham (2019)	• Short-term ethnography (in-depth semi-structured interviews with key informants, participant and non-participant observations and document analysis) • Case: Cotacachi County, Ecuador	Explores how tourism initiatives under a BV approach can lead to degrowth	BV is key to how this community conceptualizes the potentialities of tourism because it considers the wellbeing of the people and the environment. Degrowth is a consequence of BV, rather than the objective
8.		Community-owned tourism and degrowth: a case study in the Kichwa Anangu community	Renkert (2019)	• Qualitative (participant observation and semi-structured interviews with community members, ecolodge staff, and visitors) • Case: Kichwa Añangu Community, Ecuador	Explores how a project in the Kichwa Anangu community offers a compelling case study for considering how tourism could become a vehicle for developing a localized degrowth society	The sustainability of the project under study cannot be separated from its economic viability, although its success should be also measured by the contribution of tourism to a number of community-defined goals

Continued

Table 1.1. Continued

	Year	Title	Author(s)	Methodological approach	Aim	Key finding(s)
9.		Degrowing tourism: rethinking tourism	Boluk *et al.* (2019)	• Conceptual/theoretical • Unconventional narrative approach	Examines the possibility of equitable and sustainable degrowth in tourism	To overturn injustices and place tourism on a degrowth trajectory requires redefining tourism and placing it within its appropriate context of global mobilities, human wellbeing and sustainable futures
10.	2018	Degrowth in tourism: conceptual, theoretical and philosophical issues	Andriotis (2018)	Conceptual/theoretical	To explore degrowth in tourism as an alternative decommodifying philosophical approach in order to initiate a dialogue on the way that degrowth can influence travellers, residents, communities and policy makers to make responsible decisions and respect the limits to growth of each respective community	It proposes a comprehensive model explaining the transition to a degrowth tourism economy as a process following four interdependent changes that reinforce one another; namely: i) changes in behaviour; ii) focusing on locality; iii) setting limits to growth; and iv) increasing awareness. The model involves four main stakeholders (travellers, businessmen, residents and local authorities) as essential contributors to these changes

Continued

Table 1.1. Continued

	Year	Title	Author(s)	Methodological approach	Aim	Key finding(s)
11.	2014	Sustainable tourism: development, decline and degrowth: management issues from the Isle of Man	Canavan (2014)	• Mixed-method approach (postal questionnaires and in-depth personal interviews) • Case: Isle of Man, UK	To show the relevance of sustainable tourism to tourism decline and the rising interest in degrowth	Tourism decline, degrowth and economic replacement require sustainable management in order to facilitate change and various measures such as urban and rural landscape protection that may reduce adverse effects of decline
12.		Keynote speech: tourism development and the degrowth paradigm	Andriotis (2014)	Conceptual/theoretical	To look at alternatives to the classic models of development by exploring the paradigm of degrowth in a tourism context	Degrowth is revolutionary and anti-capitalistic directed to sustainable change
13.	2013	The 'antinomian' lifestyle enclave: an alternative mode of travelling	Andriotis (2013)	• Naturalistic inquiry (in-depth participant observation and in-depth interviews with travellers) • Case: island of Gavdos, Greece	To explore the crystallization of a specific antinomian travel counterculture (degrowth travellers) that advocates withdrawal from the social milieu	Antinomians have different travel orientations than that of their normative counterparts and express them in a variety of ways

Continued

Table 1.1. Continued

	Year	Title	Author(s)	Methodological approach	Aim	Key finding(s)
14.	2009	Degrowing tourism: décroissance, sustainable consumption and steady-state tourism	Hall (2009)	Theoretical/empirical	To explore the contribution of tourism to sustainable development in the context of degrowth processes and to offer an alternative discourse to the economism paradigm that reifies economic growth in terms of GDP	It suggests that considerations of sustainability must be understood within the context of political-economic discourse and proposes the concept of degrowth as a useful way of understanding the move towards a steady-state economy in which sustainable tourism is both sufficient and efficient

degrowth from different parts of the world. By doing so, it comes to explore degrowth as a strategy towards balanced tourism development and as a small, locally owned, alternative to overdevelopment and overtourism.

Following this introductory chapter which: a) identified the effects that the current COVID-19 pandemic has had on the global tourism industry and remarked that the unplanned downscaling of social and economic activity due to COVID-19 cannot be associated with degrowth; b) highlighted the differences between degrowth and overtourism and degrowth and sustainability; and c) reviewed past research on the topic of degrowth, the second chapter written by Martin Fontanari, Anastasia Traskevich and Hugues Seraphin examines the topic of overtourism to propose management solutions for destination degrowth and resilience-building. The authors use a Delphi survey, having as a sample 104 tourism experts who have explored the issue of overtourism either conceptually or empirically. These tourism experts include academics, managers of tourism associations, journalists and German ministries' representatives. By exploring their different views the study aims to understand overtourism conceptually and to develop adequate strategies to tackle this phenomenon and its negative outcomes as well as to ensure successful (de)growth.

Ernest Cañada in his contribution (Chapter 3) highlights that despite the large body of existing literature on community-based tourism there is a lack of research adopting a degrowth perspective, as well as those conditions in which degrowth can happen in the case of community-based tourism. Based on the negligence of past research, the chapter explores the potentialities and limitations of community-based tourism experiences in Central America from the perspective of a socioecological transition. From a methodological standpoint, the chapter analyses three community-based tourism initiatives in three Latin American countries: Cooperativa Los Pinos (El Salvador), Ecoposada El Tisey (Nicaragua) and Stribrawpa (Costa Rica), and highlights both their commercial success and their potential to show possible emancipatory paths. In doing so, in-depth interviews were conducted with the members of the three initiatives, and systematization of their main characteristics and results, as well as the identification of the adopted strategies, were reviewed in order to be considered as examples for a debate on how tourism can be rethought in a degrowth perspective.

In Chapter 4, Ammalia Podlaszewska undertakes an empirical analysis of tourist destinations in Indonesia to discuss some of the theoretical constructs of commoditization in community-based tourism and to explore how local resources are made available as an alternative to the dominant doctrines of 'economism'. To identify whether tourism development has exacerbated the existing forms of social and spatial equality or has provided alternative capital opportunities for the study area the author has collected data from three sources: a) the internet (newspaper articles, photos and videos about the village of Kampong Rajut); b) observations of the activities of Kampung Rajut inhabitants; and c) interviews with eight village leaders.

Chapter 5 by Neil Hughes and José Mansilla uses as a case Spanish cities such as Madrid, Palma de Mallorca, Bilbao, Seville, Valencia and Barcelona, to explore the role that degrowth social movement actors and ideas have played in protest action directed at the tourism sector in recent years. The authors identify important episodes of contestation in which degrowth activists have been present. Particularly after 2015, the Neighbourhood Assembly for Sustainable Tourism, a degrowth-inspired association made up of grassroots organizations, assemblies and groups, has made several efforts to reduce the flow of tourists to Barcelona in an attempt to reverse the damaging social, economic, cultural and environmental effects that mass tourism is having on the city. In its attempt to explore various degrowth issues, the chapter sets out a conceptual framework that draws from key literature in the field of political discourse analysis, Althusserian treatment of ideology and interpellation, and work on degrowth and tourism.

Sabine Panzer-Krause in Chapter 6 acknowledges the substantial role tour operators play in the tourism industry as intermediaries bundling different individual tourism offerings together. The study adopts an evolutionary approach through the analyses of tour operators' sustainability and audit reports and investigates whether German tour operators who have gained the CSR certification 'TourCert' have the potential to act as change agents, upscale the downscaling idea of degrowth and contribute to a reformist pathway of structural change. The findings reveal that CSR certification schemes do not seem to genuinely foster the restructuring of the tourism market within the capitalist system, but can only marginally advocate and diffuse certain elements of degrowth oriented tourism. At the same time, CSR certification schemes lack the influence necessary for a paradigm shift and for this reason the approach of degrowth-oriented tourism seems unsuitable for mainstream application.

Pascal Mayer (Chapter 7) conducted in-depth observations and interviews of locals and tourists in the Greek island of Gavdos during 2018 and 2019 in his attempt to advance the study conducted by the author (Andriotis, 2013) on antinomian travellers. He analysed the way that tourists, the majority being regulars, used to live nude under cedar trees scattered on the beaches, the so-called *kavatzas*. The study remarks that the profiles of Gavdos travellers are varied but most of them share a certain rejection of modern society, and the spaces they occupy correspond to what the French philosopher Michel Foucault called heterotopias, i.e. spaces adopting behaviours which are at odds with social rules, and everyone can feel free and in harmony with nature. According to Mayer, this profile goes beyond degrowth inspired travelling tourism; it is about the meaning of life.

Brendan Canavan's contribution (Chapter 8) identifies several negative social, cultural and environmental impacts brought about by degrowth of tourism in the Isle of Man, UK. The author points out that although the local economy has successfully transitioned to new industries, mainly offering financial services to the international wealthy and global business, the vernacular architecture, community facilities and natural landscapes have been undermined by the loss of tourist interest. Also, the sense of peripherality has been increased as tourism driven connections and exchanges have been lost. For this reason the author believes that in the case of the Isle of

Man cautions against the downsides of tourism degrowth should be taken and degrowth needs to be carefully managed if the negative consequences of degrowth are to be appreciated and mitigated.

Rasa Pranskūnienė and Dalia Perkumienė in Chapter 9 discuss the topic of freedom of movement from the point of view of tourism and degrowth. The study analyses different scientific and legal sources to give precise answers to three questions that will make it possible to rethink the meaning of freedom of movement in tourism: a) how are degrowth and freedom of movement interpreted in modern tourism?; b) how is freedom of movement interrelated with overtourism and sustainability challenges?; c) how are freedom of movement and degrowth related to the global health crisis?

The final chapter summarizes the main conclusions of the book and considers the need to rethink the phenomenon of degrowth as a new development paradigm which attempts to give up the current system of unlimited growth.

References

Alexander, S. (2012) Planned economic contraction: the emerging case for degrowth. *Environmental Politics* 21(3), 349–368. DOI: 10.1080/09644016.2012.671569.

Alexander, S. (2014) Simplicity. In: D'Alisa, G., Demaria, F. and Kallis, G. (eds) *Degrowth: A Vocabulary for a New Era*. Routledge, Abingdon, Oxfordshire, pp. 134–136.

Alexander, S. (2020) Is the economic shut down what degrowth advocates have been calling for? Available at: https://www.resilience.org/stories/2020-03-24/is-the-economic-shut-down-what-degrowth-advocates-have-been-calling-for/ (accessed 12 June 2020).

Andriotis, K. (2000) Local community perceptions of tourism as a development tool: the island of Crete. PhD thesis. Bournemouth University, Bournemouth, UK.

Andriotis, K. (2002) Options in tourism development: conscious versus conventional tourism. *Anatolia. An International Journal of Tourism and Hospitality Research* 13(1), 73–85.

Andriotis, K. (2003a) Tourism in Crete: a form of modernisation. *Current Issues in Tourism* 6(1), 23–53.

Andriotis, K. (2003b) Coastal Resorts morphology: the Cretan experience. *Tourism Recreation Research* 28(1), 67–76.

Andriotis, K. (2005) Seasonality in Crete: problem or a way of life? *Tourism Economics* 11(2), 207–227.

Andriotis, K. (2006) Hosts, guests and politics: coastal resorts morphological change. *Annals of Tourism Research* 33(4), 1079–1098.

Andriotis, K. (2008) Integrated resort development: the case of Cavo Sidero, Crete. *Journal of Sustainable Tourism* 16(4), 428–444.

Andriotis, K. (2009) Sacred site experience: a phenomenological study. *Annals of Tourism Research* 36(1), 64–84.

Andriotis, K. (2011) Genres of heritage authenticity: Denotations from a pilgrimage landscape. *Annals of Tourism Research* 38(4), 1613–1633.

Andriotis, K. (2013) The 'antinomian' lifestyle enclave: an alternative mode of travelling. *Annals of Tourism Research* 40, 40–58.

Andriotis, K. (2014) Keynote speech: tourism development and the degrowth paradigm. *In: Conference Proceedings of the 2nd Belgrade International Tourism Conference (BITSCO2014)*, College of Tourism Belgrade, Serbia 27–29 March, 2014.

Andriotis, K. (2018) *Degrowth in tourism. Conceptual, Theoretical and Philosophical Issues.* CABI, Wallingford, Oxfordshire.

Andriotis, K. and Paraskevaides, P. (forthcoming) *Negotiated exchanges in the online hospitality market: Hoteliers and Hotel managers' perceptions on booking.com.*

Andriotis, K. and Vaughan, D.R. (2004) The tourism workforce and policy: exploring the assumptions using Crete as the case study. *Current Issues in Tourism* 7(1), 66–87.

Andriotis, K. and Vaughan, D.R. (2009) The pattern-matching approach and its application in tourism development. *Current Issues in Tourism* 12(4), 315–336.

Banerjee, S.B. (2008) Corporate social responsibility: the good, the bad and the ugly. *Critical Sociology* 34(1), 51–79.

Blázquez-Salom, M., Blanco-Romero, A., Vera-Rebollo, F. and Ivars-Baidal, J. (2019) Territorial tourism planning in Spain: from boosterism to tourism degrowth? *Journal of Sustainable Tourism* 27(12), 1764–1785.

Boluk, K.A., Cavaliere, C.T. and Higgins-Desbiolles, F. (2019) A critical framework for interrogating the United nations sustainable development goals 2030 agenda in tourism. *Journal of Sustainable Tourism* 27(7), 847–864.

Bonaiuti, M. (2014) *The Great Transition*. Routledge, London.

BonJour, L. (2009) Against materialism. In: Koons, R.C. and Bealer, G. (eds) *The Waning of Materialism: New Essays*. Oxford University Press, Oxford.

Brenner, N. and Theodore, N. (2002) Cities and geographies of 'actually existing neoliberalism'. In: Brenner, N. and Theodore, N. (eds) *Spaces of Neoliberalism: Urban Restructuring in North America and Western Europe*. Blackwell, Oxford, pp. 2–32.

Brown, K. (2019) In response to exploding visitor numbers, Venice will now start charging tourists an entrance fee. Available at: https://news.artnet.com/art-world/venice-entrance-fee-tax-2019-1428708 (accessed 13 December 2019).

Buchs, M. and Koch, M. (2019) Challenges for the degrowth transition: the debate about wellbeing. *Futures* 105, 155–165.

Campbell, L.K. (2016) Getting farming on the agenda: planning, policymaking, and governance practices of urban agriculture in New York City. *Urban Forestry & Urban Greening* 19, 295–305.

Canavan, B. (2014) Sustainable tourism: development, decline and de-growth: management issues from the Isle of Man. *Journal of Sustainable Tourism* 22(1), 127–147.

Capocchi, A., Vallone, C., Amaduzzi, A. and Pierotti, M. (2020) Is 'overtourism' a new issue in tourism development or just a new term for an already known phenomenon? *Current Issues in Tourism* 23(18), 2235–2239. DOI: 10.1080/13683500.2019.1638353.

Chassagne, N. and Everingham, P. (2019) Buen Vivir: Degrowing extractivism and growing wellbeing through tourism. *Journal of Sustainable Tourism* 27(12), 1909–1925.

Cohen, E. (1973) Nomads from affluence: notes on the phenomenon of drifter-tourism. *International Journal of Comparative Sociology* 14, 89–103.

Coldwell, W. (2017) First Venice and Barcelona: now anti-tourism marches spread across Europe. *The Guardian* Thursday 10 August 2017.

Daly, H.E. (1991) *Steady-state economics*. Island Press, Washington, DC.

Demaria, F., Schneider, F., Calsamiglia, A., Blanco, L. and Domeneghini, D. (2011) Degrowth in Sudeuropa: Komplementaritat in der Vielfalt. In: Rätz, W., von Egan-Krieger, T., Muraca, B., Passadakis, A., Schmelzer, M., *et al.* (eds) *Ausgewachsen! Ökologische Gerechtigkeit. Soziale Rechte. Gutes Leben*. VSA-Verlag, Hamburg, Germany, pp. 161–172.

Fletcher, R., Mas, I.M., Blanco-Romero, A. and Blázquez-Salom, M. (2019) Tourism and degrowth: an emerging agenda for research and praxis. *Journal of Sustainable Tourism* 27(12), 1745–1763.

Fletcher, R., Mas, I.M., Blázquez-Salom, M. and Blanco-Romero, A. (2020) Tourism, degrowth, and the COVID-19 crisis. Available at: https://politicalecologynetwork.org/2020/03/24/tourism-degrowth-and-the-COVID-19-crisis/ (accessed 18 July 2020).
Fraňková, E. and Johanisová, N. (2013) Sustainable degrowth: an emerging key concept in environmental argumentation? *Socialni Studia* 10(1), 13–34.
Gascón, J. (2019) Tourism as a right: a frivolous claim against degrowth? *Journal of Sustainable Tourism* 27(12), 1825–1838.
Georgescu-Roegen, N. (1971) *The Entropy Law and the Economic Process*. Harvard University Press, Cambridge, MA.
Georgescu-Roegen, N. (1979) *Demain la Décroissance*. Pierre-Marcel Favre, Lausanne, France.
Gibson, D., Pratt, S. and Movono, A. (2013) Tribal tourism in Fiji: an application and extension of Smith's 4Hs of indigenous tourism. *Asia Pacific Journal of Tourism Research* 18(8), 894–912.
Gössling, S., Scott, D. and Hall, C.M. (2020) Pandemics, tourism and global change: a rapid assessment of COVID-19. *Journal of Sustainable Tourism* 12(5), 1–20. DOI: 10.1080/09669582.2020.1758708.
Hall, C.M. (2009) Degrowing tourism: décroissance, sustainable consumption and steady-state tourism. *Anatolia: An International Journal of Tourism and Hospitality Research* 20(1), 46–61.
Hall, C.M. (2010) Changing paradigms and global change: from sustainable to steady-state tourism. *Tourism Recreation Research* 35(2), 131–145.
Hall, C.M. (2011) Consumerism, tourism and voluntary simplicity: we all have to consume, but do we really have to travel so much to be happy? *Tourism Recreation Research* 36(3), 298–303.
Higgins-Desbiolles, F. (2018) Sustainable tourism: sustaining tourism or something more? *Tourism Management Perspectives* 25, 157–160.
Higgins-Desbiolles, F. (2020) Socialising tourism for social and ecological justice after COVID-19. *Tourism Geographies* 22(3), 610–623. DOI: 10.1080/14616688.2020.1757748.
Hinrichsen, D. (1987) *Our Common Future: A Reader's Guide*. Earthscan, London.
Hunter, C. (1997) Sustainable tourism as an adaptive paradigm. *Annals of Tourism Research* 24(4), 850–867.
Illich, I. (1973) *Tools for Conviviality*. Calder and Boyars, London.
J.P. Morgan Chase and Co (2020) What is the global impact from the coronavirus outbreak? Available at: https://www.jpmorgan.com/global/research/coronavirus-impact (accessed 20 June 2020).
Janssen, H., Kiers, M. and Nijkamp, P. (1995) Private and public development strategies for sustainable tourism development of island economies. In: Coccossis, H. and Nijkamp, P. (eds) *Sustainable Tourism Development*. Avebury, London, pp. 65–83.
Johnsen, C.G., Nelund, M., Olaison, L. and Meier Sørensen, B. (2017) Organizing for the Post-growth economy. *Ephemera: Theory and Politics in Organization* 17(1), 1–21.
Jonas, A.E.G. and Gibbs, D.C. (2003) Changing local modes of economic and environmental governance in England: a tale of two areas. *Social Science Quarterly* 84, 1018–1037.
Kallis, G. (2011) In defence of degrowth. *Ecological Economics* 70, 873–880.
Klitgaard, K.A. and Krall, L. (2012) Ecological economics, degrowth, and institutional change. *Ecological Economics* 84, 247–253.
Kousis, M. (2000) Tourism and the environment: a social movements perspective. *Annals of Tourism Research* 27(2), 468–489.
Laing, J.H. (2006) Extraordinary journeys: motivations behind frontier travel experiences and implications for tourism marketing. PhD thesis. La Trobe University, Melbourne, Australia.
Laing, J.H. and Crouch, G.I. (2009a) Lone wolves? Isolation and solitude within the frontier travel experience. *Human Geography/Geografiska Annaler B* 91(4), 325–342.

Laing, J.H. and Crouch, G.I. (2009b) Myth, adventure and fantasy at the frontier: metaphors and imagery behind an extraordinary travel experience. *International Journal of Tourism Research* 11(2), 127–141.

Latouche, S. (2006) The globe downshifted: how do we learn to want less? *Le Monde Diplomatique, January*.

Latouche, S. (2009) *Farewell to Growth*. Polity Press, Cambridge.

Latouche, S. (2010) Growing a degrowth movement. In: WorldwatchInstitute (ed.) *State of the World 2010: Transforming Cultures from Consumerism to Sustainability*. W.W. Norton & Company, New York, p. 181.

MacCannell, D. (1992) *The Empty Meeting Grounds*. Routledge, New York.

Mangili, A. and Gendreau, M.A. (2005) Transmission of infectious diseases during commercial air travel. *Lancet* 365(9463), 989–996. DOI: 10.1016/S0140-6736(05)71089-8.

Martin, D.G. (1998) Generational differences amongst new age travellers. *Sociological Review* 46(4), 735–756.

Martin, D.G. (2000) Constructing place: cultural hegemonies and media images of an inner-city neighborhood. *Urban Geography* 21, 380–405.

Marx, K. (1859) *A Contribution to the Critique of Political Economy*. Progress Publishers, Moscow.

Meadows, D.H., Meadows, D.L., Randers, J. and Behrens, W.W. (1972) *Limits to Growth*. Universe Books, New York.

Milano, C., Novelli, M. and Cheer, J. (2019a) Overtourism and degrowth: a social movements perspective. *Journal of Sustainable Tourism* 27(12), 1857–1875.

Milano, C., Novelli, M. and Cheer, J.M. (2019b) Overtourism and tourismphobia: a journey through four decades of tourism development, planning and local concerns. *Tourism Planning and Development* 16(4), 353–357.

Monterrubio, J.C. (2010) [illegible] economic impacts of influenza A (H1N1) and government reaction on the Mexican tourism industry: an analysis of the media. *International Journal of Tourism Policy* 3(1), 1–15.

Navarro-Jurado, E., Romero-Padilla, Y., Romero-Martínez, J.M., Serrano-Muñoz, E., Habegger, S. *et al.* (2019) Growth machines and social movements in mature tourist destinations Costa del Sol-Málaga. *Journal of Sustainable Tourism* 27(12), 1786–1803.

Nelson, A. and Liegey, V. (2020) Coronavirus and degrowth. *Ecologist* 3rd April.

Owen, R.E., Witt, S. and Gammon, S. (1993) Sustainable tourism development in Wales: from theory to practice. *Tourism Management* 14(3), 463–474.

Panzer-Krause, S. (2019) Networking towards sustainable tourism: innovations between green growth and degrowth strategies. *Regional Studies* 53(7), 927–938.

Pappalepore, I., Maitland, R. and Smith, A. (2014) Prosuming creative urban areas. Evidence from East London. *Annals of Tourism Research* 44, 227–240.

Perles-Ribes, J.F., Ramón-Rodríguez, A., Vera-Rebollo, J.F. and Ivars-Baidal, J. (2018) The end of growth in residential tourism destinations: steady state or sustainable development? The case of Calpe. *Current Issues in Tourism* 21(12), 1355–1385.

Reichel, A. (2015) re-taking sustainable development for degrowth. Available at: https://www.degrowth.info/en/2015/12/re-taking-sustainable-development-for-degrowth/ (accessed 2 June 2018).

Renkert, S.R. (2019) Community-owned tourism and degrowth: a case study in the Kichwa Añangu community. *Journal of Sustainable Tourism* 27(12), 1893–1908.

Schneider, F., Kallis, G. and Martinez-Alier, J. (2010) Crisis or opportunity? Economic degrowth for social equity and ecological sustainability: introduction to the special issue. *Journal of Cleaner Production* 18(6), 511–518.

Schumacher, E.F. (1973) *Small is Beautiful*. Penguin, London.

Seraphin, H., Gladkikh, T. and Vo Thanh, T. (2020) *Overtourism Causes, Implications and Solutions*. Palgrave, New York.

Sharpley, R. (2002) Tourism: a vehicle for development? In: Sharpley, R. and Telfer, D.J. (eds) *Tourism and Development: Concepts and Issues*. Channel View Publications, Clevedon, UK, pp. 11–34.

Smith, M., Macleod, N. and Robertson, M.H. (2010) *Post-Tourism*. Sage, London.

Stroebel, M. (2015) Tourism and the green economy: Inspiring or averting change? *Third World Quarterly* 36(12), 2225–2243.

Tapper, M.L. (2006) Emerging viral diseases and infectious disease risks. *Haemophilia* 12 Suppl 1(1), 3–7. DOI: 10.1111/j.1365-2516.2006.01194.x.

The Guardian (2020) Cruise ship accounts for more than half of virus cases outside China: as it happened. 20 February 2020.

Tukker, A. (2014) Strategies for enhancing resource efficiency. In: Angrick, M., Burger, A. and Lehmann, H. (eds) *Factor X. Eco-Efficiency in Industry and Science*. 29. Springer, Dordrecht.

UNWTO (2019) *International Tourism Highlights 2019 Edition*. World Tourism Organization, Madrid.

UNWTO (2020) International tourist arrivals could fall by 60-80% in 2020. Available at: https://www.unwto.org/news/COVID-19-international-tourist-numbers-could-fall-60-80-in-2020 (accessed 22 June 2020).

Valdivielso, J. and Moranta, J. (2019) The social construction of the tourism degrowth discourse in the Balearic islands. *Journal of Sustainable Tourism* 27(12), 1876–1892.

Vogt, J.W. (1976) Wandering: youth and travel behavior. *Annals of Tourism Research* 4(1), 25–41.

WCED (1987) *Our Common Future*. World Commission on Environment and Development. Oxford University Press, Oxford.

Weiss, M. and Cattaneo, C. (2017) Degrowth taking stock and reviewing an emerging academic paradigm. *Ecological Economics* 137, 220–230. DOI: 10.1016/j.ecolecon.2017.01.014.

WTTC (2020) WTTC press release. Available at: https://www.wttc.org/about/media-centre/press-releases/press-releases/2020/coronavirus-puts-up-to-50-million-travel-and-tourism-jobs-at-risk-says-wttc/ (accessed 2 June 2020).

2 (De)growth Imperative: The Importance of Destination Resilience in the Context of Overtourism

Martin Fontanari[1]*, Anastasia Traskevich[2] and Hugues Seraphin[3]

[1]ISM International School of Management GmbH, Cologne, Germany
[2]Belarusian State Economic University, Germany
[3]University of Winchester, Winchester, UK

2.1 Introduction

Since the early 1970s, criticism of the tourism industry for its excessive volume and negative impact on the local environment has become regular and traditional. Overtourism, however, has been addressed as a more recent phenomenon, and denotes the 'touristification' of a habitat, including all negative economic, ecological and social outgrowths that such an excess of tourism entails (Oklevik *et al.*, 2019; Panayiotopoulos and Pisano, 2019; Wall, 2019). However, the current research is focused not on the quantitative effects of overtourism, which are narrowed down by the COVID-19 pandemic. Rather, the investigation concerns the fact of 'touristification of a habitat' and all the related negative economic, ecological and social impacts of tourism in regions and for communities – the effects of which comprise a pressing issue even, or especially, in times of systemic crisis caused by the pandemic.

The devastating effects of COVID-19 bring a new understanding to the topic of overtourism. The issue cannot be considered 'dead' through the current disappearance of crowds in touristic hotspots. In the current situation, the negative effects of overtourism to the overall economic, social and cultural orientation of a region can be perceived as even acuter than ever: destinations find themselves unable to react adequately to challenges, no matter whether these may be the excessive numbers of tourists or no tourists

*Corresponding author: martin.fontanari@ism.de

DOI: 10.1079/9781789245073.0002

at all. The initial and native economic and social orientation of a region and its community has been replaced by global economic structures which aim at the development of mass tourism in the destination. This means that in critical situations, like a pandemic outbreak, such destinations appear to be totally locked down and unable to come back to their initial authentic way of life and economic activities. The first attempts to reflect on the current situation come to the conclusion that more holistic strategic solutions are required to balance tourism development in destinations which were once affected by overtourism.

In the context of the present critical situation of regions previously affected by overtourism, current research aims to reflect on the phenomenon of overtourism conceptually – to develop adequate strategies to balance tourism development in regions in 'the new future' (Zukunftsinstitut, 2020) created by COVID-19.

At the current stage of research, some attempts have been made to develop initial solutions to the problem of overtourism (e.g. Haifeng *et al.*, 2012; Muler Gonzalez *et al.*, 2018; Dodds and Butler, 2019; Gretzel, 2019). However, these efforts did not lead to any holistic breakthroughs, even though a certain shift away from purely quantitative tourism concepts is apparent worldwide. For urban and rural destinations which once suffered from overtourism, new design approaches and measuring instruments should be taken into consideration, in order to better understand the phenomenon of tourism growth as such, and to draw strategic conclusions for further destination development. This may lead to overcoming the still pressing negative effects of overtourism in a holistic way.

The study which was launched at the beginning of 2019 took as its objective that of finding a consensus between the different suggested strategies to balance tourism growth and tackle the related perverse impacts of overtourism. This led to two core research questions:

- What are the different views (from the perspective of a stakeholder) regarding the strategies which need to be put in place to tackle the effects of overtourism and related perverse impacts?
- What could be a potential strategic consensus to tackle the effects of overtourism and ensure successful (de)growth?

In these terms, (de)growth as a strategic consideration is valued as a tool for development of qualitative aspects of tourism dedicated to discovery and preservation of uniqueness, authenticity and identity of destinations. For this purpose, the theory of resilience is applied in the current study to emphasize the safeguarding and balancing functions of the tourism industry, which become especially topical within the present devastating effects of the pandemic. The first attempts to analyse the implications of the current pandemic showed that the present situation of a systemic crisis required thorough investigation within the theory of resilience (Cheema-Fox *et al.*, 2020; Fontanari and Traskevich, 2020; Gössling *et al.*, 2020; Hall *et al.*, 2020; Lapointe, 2020; Nepal, 2020; Niewiadomski, 2020).

From a methodological point of view, this study began by collecting data from experts in the field of tourism development and, more specifically, people who have investigated the impacts of overtourism either conceptually or empirically (Fontanari and Berger-Risthaus, 2020). As a basis for the later Delphi survey and its underlying theses and statements, an extensive literature-based search into the causes and justifications of overtourism was first conducted (Milano, 2017; Moscardo *et al.*, 2017; Koens *et al.*, 2018; Jacobsen *et al.*, 2019). More than 100 authors who have devoted themselves to the problem of overtourism in public and scientific articles were identified, although this list does not claim to be exhaustive and centres on German and English academic literature. From this literature analysis, 28 theses and statements were proposed to conduct a Delphi survey (Innerhofer *et al.*, 2020).

The expected outcome of this contribution was to offer a practical strategic approach to destination management in order to tackle the effects of overtourism for any case of physical quantities of tourists in destinations (up to none in the situation of a pandemic) through the following theoretical approaches:

- the idea of (de)growth strategies which allow balancing of tourism development and focus it on unique and authentic touristic experiences; and
- the resilient orientation of destinations to ensure their safeguarding functions in any scenario of economic, social, cultural, ecological and political development in the regions.

2.2 Methodology of Empirical Study

The Delphi method assesses certain development processes in order to allow for a look into the future, thereby accelerating the implementation of innovations. In the literature there is no uniform procedure for the use of the Delphi method, which permits a broad spectrum of survey types and variation of them. Results also serve as the information basis for decision making (Blind and Cuhls, 2001, p. 59). In a Delphi survey, it is not the statistical representativeness which is relevant, but rather the expert knowledge which must be recorded and derived with the intention of evaluating the current state of the art and future forecasts (Niederberger and Renn, 2018). The Delphi method is used to produce oracular statements regarding the future. This method requires the assembling of a panel of experts with a wide range of experience (Hammond and Wellington, 2013).

2.2.1 Adaptation of the Delphi method within the Welphi platform

Within the 28 hypothetical statements, a three-stage Delphi survey was conducted in the form of an online questionnaire, with a mix of qualitative and quantitative statements. For this purpose, the specialized online survey

Table 2.1. Survey panel.

	1st round			2nd round			3rd round		
Expert group	N	n	Rate	N	n	Rate	N	n	Rate
Academics	34	13	38%	34	13	38%	13	12	92%
Managers (tourism associations)	23	1	4%	23	4	17%	4	4	100%
Journalists	25	6	24%	25	1	4%	6	2	33%
Ministries (Germany)	22	2	9%	22	2	9%	2	2	100%
Others	-	3	-	-	2	-	3	1	33%
Total	**104**	**25**	**24%**	**104**	**22**	**21%**	**28**	**21**	**75%**

Note: N – the number of experts invited for the round of Delphi survey; n – the number of experts who participated in the round of the survey.
Source: The authors

platform Welphi (www.welphi.com) was used, which allowed an easy and affordable way for the questionnaires to canvass the opinions of 104 participants, across a geographically dispersed area and with busy agendas. The range of experts involved is presented in Table 2.1.

Through a system of three rounds, it is possible to confront each participant anonymously with the opinions of his or her peers, engaging participants in a non-face-to-face format and promoting consensus across the group on issues that lack the support of quantitative data.

Thus, in terms of methodology, the Welphi approach ranges from a face-to-face Delphi meeting, where the confrontation of ideas is possible, but the 'physical' presence of all participants is required, to a simple online questionnaire, where the opinion of participants is requested but not shared with others and no opportunity is given to each participant to revise his or her opinion under the influence of the opinions of peers. Therefore, the Welphi approach allows for the confrontation of ideas in an asynchronous, online, participatory and interactive way. Moreover, by allowing the sharing of participants' comments anonymously, each participant can change their initial position, thus promoting group consensus.

2.2.2 Research approach to quantitative evaluation of the statements

The present research approach presented a quantitative evaluation of the diverse data, which was received within the statements. The statements are initially arranged on an ordinal scale of 1–5 in the first round; this scale is standardized for all the questions and ranges from '1 = completely disagree' to '5 = fully agree'. At the same time, the experts were given the opportunity to reveal qualitative positions on the individual statements and to incorporate different arguments.

In the first round, the experts were able to give their opinion on a total of 28 statements and comment on them with the help of the commentary function. Following the first round of the survey, it was possible to select the statements which could be evaluated as already sufficiently answered. The other statements, which had the same formulation of the question, were taken over to the second round and, in the same way, to the third round, in order to achieve a clearer expert opinion. With the further rounds, the approval scale was reduced according to the Delphi methodology: from 1–5 to the three most frequent responses (for the second round) and the two most frequent responses (for the third round). The scale values were reassessed accordingly.

To explain this more precisely, the methodological decisions to remove or adopt statements from the Delphi survey and reassess the scale values were the following:

- For the first round: if a particular scale value results in an expert approval rate of 50% or more; and, in the adjoining consent value (e.g. 'agree' + 'rather agree'), if the rate is more than 25%, such a question is regarded as meaningful; further expert discussion is then not necessary.
- For the second round: expert result is considered meaningful if the scale value results in the approval rate of 51% or more; further expert discussion is then not necessary.
- If the approval rate is 50% or less, the statement is included in the second (third) round for further expert discussion.
- If a statement received too much scope in experts' interpretation and thus, too many varied approval rates, it is not adopted for the second round.

The evaluation and display of the results are as follows:

1. Presentation of the average agreement, on the basis of a semantic profile.
2. Naming of the statement and representation of the individual scale values, agreement rate in %.
3. Analysis of the comments made by the experts.
4. The researchers' decision regarding the continuation or removal of a statement.
5. If a statement is included in the second (third) Delphi round, the underlying three- (two) part evaluation scale is also explained for each statement.

For this chapter, the results of the Delphi study were selected to emphasize the idea of degrowth strategies for tourism destinations. At the same time, within the selected statements, the benefits of resilient orientation of destinations (Fontanari and Traskevich, 2020) are shown in the context of managing overtourism and implementing degrowth strategies.

2.3 Main Findings

In this section, selected statements are presented from 28 questions which were offered to the experts (Table 2.2). Each of the selected questions is illustrated with the statistical results on the rates of experts' approval in each

Table 2.2. Selected results of the Delphi survey on overtourism.

Statement 1	The issue of overtourism should not be viewed critically as a problem in the medium term and will be resolved automatically in the future.						
Approval	**5**	**4**	**3**	**2**	**1**	**Mean**	**Comments**
Round 1	4%	11%	4%	26%	56%	1.84	2
Round 2	No further round necessary						
Statement 2	Overtourism is a phenomenon that is perceived and felt more sensitively by the locals with increasing prosperity in a region/destination.						
Approval	**5**	**4**	**3**	**2**	**1**	**Mean**	**Comments**
Round 1	23%	54%	19%	4%	0%	3.96	0
Round 2	36%	55%	9%			4.27	3
Round 3	No further round necessary						
Statement 3	Overtourism is a question of individual perceptions, both for locals and for tourists.						
Approval	**5**	**4**	**3**	**2**	**1**	**Mean**	**Comments**
Round 1	15%	27%	27%	15%	15%	3.12	2
Round 2	No further round necessary						
Statement 4	The declining quality of life of the locals due to overtourism is clearly noticeable and measurable.						
Approval	**5**	**4**	**3**	**2**	**1**	**Mean**	**Comments**
Round 1	46%	19%	15%	15%	4%	3.88	1
Round 2	27%	59%	14%			4.13	3
Round 3	No further round necessary						
Statement 5	Overtourism results from the accumulation of non-existent strategic planning and the absence of spatial and destination design.						
Approval	**5**	**4**	**3**	**2**	**1**	**Mean**	**Comments**
Round 1	26%	33%	30%	4%	7%	3.67	2
Round 2	35%	43%	22%			4.13	1
Round 3	33%	67%				4.33	4
Statement 6	In order to achieve the highest possible benefit for residents and tourists, urban and local development concepts are approached jointly by tourism managers and urban planners. The focus is on participatory processes from idea development to decision making in the areas of infrastructure, environment and quality of life.						
Approval	**5**	**4**	**3**	**2**	**1**	**Mean**	**Comments**
Round 1	15%	27%	27%	15%	15%	3.12	2
Round 2	59%	27%	14%			4.45	3
Round 3	No further round necessary						
Statement 7	In developing a future strategy for a destination, the various stakeholder groups must be involved, whether or not they are directly linked to tourism. Tourism development cannot be left to tourism professionals.						
Approval	**5**	**4**	**3**	**2**	**1**	**Mean**	**Comments**
Round 1	65%	31%	4%			4.61	2
Round 2	No further round necessary						

Continued

Table 2.2. Continued

Statement 8	The affected destination focuses on the development of tourism in the surrounding area and on more effective domestic marketing, and creates a relief of its hot-spots through attractive products in the low season.						
Approval	**5**	**4**	**3**	**2**	**1**	**Mean**	**Comments**
Round 1	12%	31%	38%	12%	8%	3.27	3
Round 2		41%	27%	32%		3.09	4
Round 3		57%	43%			3.57	4
Statement 9	Communicating the positive aspects of tourism in general to the inhabitants is sufficient to counteract overtourism. Justify your response in the comment function on the right.						
Approval	**5**	**4**	**3**	**2**	**1**	**Mean**	**Comments**
Round 1	12%	4%	4%	38%	42%	2.06	17
Round 2			5%	23%	73%	1.30	11
Round 3	No further round necessary						
Statement 10:	In order to sensitize each individual stakeholder to the topic, there will be organized training courses and e-learning platforms organized by the respective destination management organization (DMO), which will increase and certify the quality of consulting and services offered by the service providers, in terms of visitor flow management.						
Approval	**5**	**4**	**3**	**2**	**1**	**Mean**	**Comments**
Round 1	16%	20%	48%	16%		3.36	
Round 2	No further round necessary						

Note: the approval scale values stand for the following: 5 – completely agree; 4 – rather agree; 3 – partial/ partial; 2 – rather disagree; 1 – completely disagree.
Source: The authors

round. Also, the number of experts' additional comments is indicated, and the mean scale value is calculated for each response (see Section 2.2 of this chapter).

The analysis shows the importance of addressing the issue of quality of life in tourism destinations, as well as the importance of adopting a new comprehensive system of destination management which would incorporate the concepts of degrowth and resilience. Selected expert comments were analysed and interpreted for further conceptual considerations and managerial suggestions of the study.

Statement 1 (Table 2.2) was about the temporary nature of the very issue of overtourism. Already within the first round, a clear picture was received: 82% of the experts (rather) denied this statement, with the mean value of the rejection 1.84.

According to the responses given, it was clear that, even if one assumes that overtourism is just a short-term problem, it is still one ' … that needs to be managed'. Tourism 'requires a sustainable concept', since ' … the natural heritage does not need tourism, but tourism is dependent on the natural heritage'. It can therefore be concluded that the problem of overtourism will not be solved by itself in the future. Rather, overtourism requires new concepts of dealing with this phenomenon, in order to protect the entire tourism system.

Furthermore, more than 90% of the experts (rather) agreed that overtourism is a phenomenon that is perceived and felt more sensitively by the locals, even in cases in which their wealth increased over time with the impact of tourism (Table 2.2).

Experts mention that the intensity of perceived overtourism depends on the target group as well as the very nature of the destination. Obviously, there are differences between rural and urban areas. In many areas, tourism is an important economic branch, through which service providers and locals can support their families. In this situation, all other relevant objectives (e.g. environmental or cultural protection) are rejected in favour of earning an income. An expert statement illustrates the main problem: 'Especially in worse managed regions, tourism passes by the local population, which has to bear negative effects.' There might be two conclusions from this:

- classic, established destinations are more sensitive to overtourism on account of their social and cultural nature
- on the other hand, the sensitivity to overtourism depends on the quality of the destination management.

In 'just growth-oriented' destinations, the positive impacts of tourism don't actually benefit the local population due to high leakage. At the same time, the natives have to endure the negative effects.

The most challenging topic of overtourism is how to measure or methodically capture this phenomenon. Neither natives nor tourists naturally rely on any actual model of measurement. Furthermore, the impression of being affected by overtourism is also heavily influenced by peer groups and media. Statement 3 (Table 2.2) of the Delphi study aims at this individually distinctive perception of the phenomenon by approaching both these aspects. The challenging nature of this question is revealed in the very first round when a whole range of expert opinion was revealed. Even the distribution of the experts' opinions – mean value achieved at 3.12 – does not methodologically allow for the sorting-out of further scale values.

However, this 'unsolved' statement remains one of the most crucial aspects in approaching overtourism with articulate solutions. While many of the experts confirm the importance of the phenomenon of 'individual perception' (Muler Gonzalez *et al.*, 2018), others believe that existing territory-oriented instruments for assessment-carrying capacities of the recreational landscapes are still sufficient to indicate the impact of overtourism, even with all its current social and cultural facets. Even if overtourism is a matter of perception, this does not exclude the possibility that it might be statistically or methodically measured as a real-life problem. Therefore, this statement still appears relevant for the derivation of management approaches, even though subjective perceptions of the problem of overtourism by travellers or locals would require extensive quantitative and qualitative surveys and further measuring models.

Statement 4 (Table 2.2) was about the measurable quality of life of the locals due to overtourism. The sharp deviation of the responses and comments can be connected with the difficulty of measuring both the quality of life and the impact of overtourism.

Some experts pointed out that the perception to overtourism is subjective and depends on how much one is exposed to overcrowding. The problem is based on the questions of how to define, measure and compare quality of life in the situation when every city or region has very different conditions. What all expert statements have in common, however, is that the assessment and evaluation of limited quality of life is based on subjective, and thus different, impressions and requires a more objective – and so more systematic – approach and measurement. The experts, therefore, express the need to define and quantify related impacts – but are aware that there is no current instrumental set to correlate local quality of life with the impact of overtourism.

The next statement was designed around the hypothesis that the perception of overtourism by locals and guests also results in a lack of involvement or consideration of their interests and expectations for the strategic development of a destination (Statement 5, Table 2.2). Above all, this statement assumes that the requirement of strategic and integrative planning in the context of holistic destination development at the current moment often fails to meet high theoretical and methodological standards when implemented in real practice. The growth of the mean value over the rounds shows clearly that the experts tended to crystallize their opinion regarding the special meaning of relevant strategic planning, in order to cope with both hard and soft factors of overtourism.

Analyses of the comments of the experts provides an explanation of the still cautious agreement with this statement (67% rather agree). The experts mention the importance of strategic planning, but see a number of circumstances and influencing factors which prevent the desired success in real practice.

Some experts first point out the particular significance of strategic planning and a long-term vision: 'If a city does not strategically know what it wants to be and how many visitors it can accommodate, it will overflow and lose control. What is needed is long-term strategies and visions which are currently not popular with city administrations.'

At the same time, the statements of many experts also explain why strategic planning, nevertheless, often fails. In particular, numerous unpredictable sociocultural factors still expose the destination to overtourism, even when the planning was successfully implemented. Some successful tourism planning is often not devoted to holistic sustainable regional development in the sense of receiving welfare and economic growth for all the residents and market participants. Hence, experts claim: ' … the question arises how the positive aspects of tourism can affect urban development and how it can be exploited holistically'. Moreover, if strategic tourism planning is successfully implemented, but only to the benefit of the tourism industry and its stakeholders, not to the holistic welfare of the local community, this can still lead to overtourism. As the experts wrote: 'also the planning is often incoherent and does not apply to cross-sector integration processes'. In all these complex cases, overtourism is not really the result of a lack of strategic urban planning itself, but rather a result of the lack of holistic understanding of the strategic significance of tourism to the region and its community.

Among such problematic social and cultural factors, experts also mention ' ... a dramatic increase in the volume of the number of international tourists – which cannot just be solved by better cooperation and planning'. A lack of ongoing professionalization of tourism and destination management is also mentioned as a home-made reason for overtourism: 'they have lived from hand to mouth for too long, with one-sided, short-term objectives; strategic management skills have been too modestly represented in too many decision-making bodies for too long'. On the basis of these statements, the following conclusions can be made:

- strategic planning for destination development should become integrated and oriented to the long term (for a planning and design period of 15–20 years)
- it should take place in a cross-sectoral mode
- the quality of management competencies in decision-making bodies should be increased; and
- a new methodological approach and conceptual understanding of tourism development is required, which will address the risk of overtourism in detail.

The next statement under critical evaluation within the Delphi study was about the cooperative approach of urban planners and tourism managers (Statement 6, Table 2.2). After the second round of questioning, there was clear agreement with this statement, with a mean of 4.45 and almost 60% agreement.

Almost all experts relied on a cooperative approach in tourism planning and development, but the slowness of or restrictions by the administration is often a challenge here. When the framework of planning is extended because of inviting other specialists from other planning departments, the execution of the planning process becomes highly complicated. As the experts say, 'having the same database, looking together at a city and its development can improve the situation in the destination. The big steering approaches cannot be implemented without the urban planning and the legal authorities of the city or even the national government ... '. Experts are aware here that an integrative planning process is ideal to achieve the goal of holistic planning. However, at the same time this is very difficult to implement effectively, when so many stakeholders and special forces are actually involved. 'Administration and local council have, additionally, a very strong perseverance which too often causes no future-oriented development to take place.'

These statements can be interpreted as follows:

- integrative and participatory development planning is the key strategic tool to handle the problem of overtourism;

- tourism management and urban management both need new approaches to implement participative integrative decision-making procedures.

Furthermore, experts are agreed on the fact that 'tourism development cannot be solely left to tourism experts'. Statement 7 (Table 2.2) points out that every destination requires synergy-oriented planning. Two thirds of the experts agreed on that idea, even within the first round.

Experts also commented that various stakeholder groups which are not directly involved in tourism will use their external insights to fertilize professional decisions in tourism development. Aspects for increasing the quality of leisure time, and thus the overall quality of life of locals, also influence the attractiveness of business locations. One can assume that external stakeholders or entrepreneurs who do not come directly from the tourism industry can enrich tourism planning with new ideas.

It is especially important to ensure that within tourism planning synergies are fulfilled and complementary gaps are covered to contribute tangibly to the quality of life of both guests and locals. This may be achieved through meticulous cooperation-based project moderation. Synergetic and complementary considerations can each be taken into account in the form of benefit calculations.

At the same time, effective approaches for holistic planning would also include geographical and spatial components towards the product design (Statement 8, Table 2.2).

Some 57% of the respondents rather agree with the statement that the overcrowded destination should concentrate on a triad of focused activities:

- to extend tourism into the surrounding geographical areas;
- to elaborate on effective domestic marketing; and
- to design attractive products for the off-season.

However, experts comment about their own experience in dealing with seasonality in tourism in real-life-conditions (e.g. weather in the off-season, school or holiday periods, etc.). 'Off-season trips are off-season because the weather is usually less attractive and does not coincide with school holidays.'

It is even more difficult to implement a divergent geographical approach towards spatial tourism design and planning: ' … that's not how we travel. The masses attract the masses. We mainly travel to the well-known areas and attractions'. The reason for this is that the tourists want to see the hotspots or 'need' the top-rated spots for a selfie or Instagram photo. Only a limited number of travellers leave the beaten track.

Moreover, strengthening the low season does not often lead to a 'smoothing' of the high season. In fact, such efforts of product design can also attract new guests (e.g. Jin and Sparks, 2017), so that 'the low seasons also become high seasons'. The increased efforts to attract guests in the low season, therefore, lead to selecting and addressing new target groups, 'who show the desired behaviour'. 'Cities in particular attract a global audience, and every city must define its ideal target groups in comparison with all players of the visitor economy.' Experts state, however, that the classic destinations affected by overtourism have in the meanwhile become hotspots throughout the year (e.g. Barcelona, Venice, Amsterdam).

With the same kind of reasoning, some experts also deny that activating potentials in the surrounding area of an overcrowded site will lead to relief. In contrast, 'by realizing tourist attractions in the surroundings, you simply make the destination more attractive, not less attractive'. Besides this, experts mention the existing 'ivory tower thinking, municipal district boundaries and public tourism promotion oriented to political-administrative borders'. These factors do not allow for the spatial expansion of the portfolio of a destination. Experts agree on creating new points of attraction to expand carrying capacities and seasonal distribution. But stronger domestic marketing is also required and can be seen as a much more differentiated solution. The context of internal marketing, and the possibilities of a more focused 'internal communication' with service providers and locals for the communicative control of (regular) guests, has, however, not been sufficiently brought up for discussion in the statements.

The most intensively and most controversially discussed thesis was Statement 9, to which additional qualitative statements were also expressly invited (Table 2.2).

In the second round, 73% of experts completely disagreed with this statement (with a mean of 1.3). However, the comments submitted by the experts for it are diverse and ground-breaking at the same time. First of all, experts mention the negative aspects of high-density tourism, such as 'emissions, housing shortage, increased cost of living and rents, low remuneration in service-dominant tourism ... which cannot be corrected simply by explaining the positive aspects (tourism value added, jobs, etc.)'. Moreover, 'in some overcrowded areas, such as cities where locals and tourists share space, locals tend to perceive more negative than positive aspects of tourism ... , so the positive effects of tourism hardly represent a counterbalance'. Further: 'Simply communicating the positive effects of tourism as an economic factor is not enough to appease critics and does not automatically lead to a solution to the problem.'

Experts also stated that ' ... the feeling for overtourism often comes from the gut, while the argumentation aims at the head – which causes different levels of sensation and communication'. Besides, in their comments, they reduce to a certain level the importance of open communication. Yet while this activity helps to understand the effects of tourism, it is by no means the only and sufficient activity.

The essence of these contradictory points of view lies in the fact that 'what is economically attractive for a city or region is not necessarily the same as what makes a city/neighbourhood liveable and lovable for its inhabitants. These qualities often compete with each other because the needs are different'.

However, it can be assumed that this disagreement brings one back to the statements on participative integrative planning (6 and 7, Table 2.2). This requires strategic planning to be in line with both residents' and guests' needs: 'Building strong partnerships and cooperation is the most important basic element of management of overtourism. In addition, the general positive attitude of the locals towards the guests is necessary. Both imply the conviction of the positive importance of tourism development in the region.'

This aspect requires that the responsible sides both in tourism and city management have to elaborate a more intensive and differentiated method of communication between all the involved people and stakeholders. The determinants to be shaped within such strategic communication are regional identity, homeland, solidarity, value reference, cultural reference, tradition and historic memory, among others. Certainly, such determinants are only created with the tools which go far beyond classical marketing communication in tourism (e.g. Park and Petrick, 2006). This requires new perspectives and a new quality of communication within 'interior marketing' towards 'resilient identity' (Traskevich and Fontanari, 2018). Related to this, the experts have also been invited to discuss the statement which presented some future instruments for both better interior marketing and competence for improved visitor-flow management, such as e-learning-platforms or organized training courses (Statement 10, Table 2.2).

The cautious approval showed by the experts can be explained by the fact that the examples mentioned in the statement are only a few of the many possible ones. At any rate, it can be seen in the trend that new instruments and measures are needed to raise the awareness of service providers and destination management, and to set stronger emphasis on the internal marketing of a destination for at least a balanced consideration of the phenomenon of overtourism.

2.4 Supplementary Tourism Satellite Accounts Address the Problem of Overtourism

It should first be noted that the phenomenon of overtourism as such is not new, but has clear parallels with the problems of mass tourism and the subsequent 'revolt of the travelled' (such as in India in the 1970s). Accordingly, approaches and methods of destination (regional) planning, which can be integrated to facilitate the better avoidance or overcoming of critical tourism developments or negative effects, already exist.

The Delphi study came up with quite a clear understanding of the view of the experts, stating that the overtourism phenomenon not only presents itself as a problem in large cities, but also leads to an increasing impairment of the quality of life of some rural regions, such as the Alps, coastal regions or UNESCO World Heritage sites.

Since there are no models or measuring methods which express the objective and subjective perceptions of locals about overtourism and its associated effects on quality of life, this must be the starting point for further research in tourism science. It should also be noted that the subjective perception of stress and impairment can only be ascertained with difficulty. Thus, it is relevant also for the question of a representative, measurable quality of life,

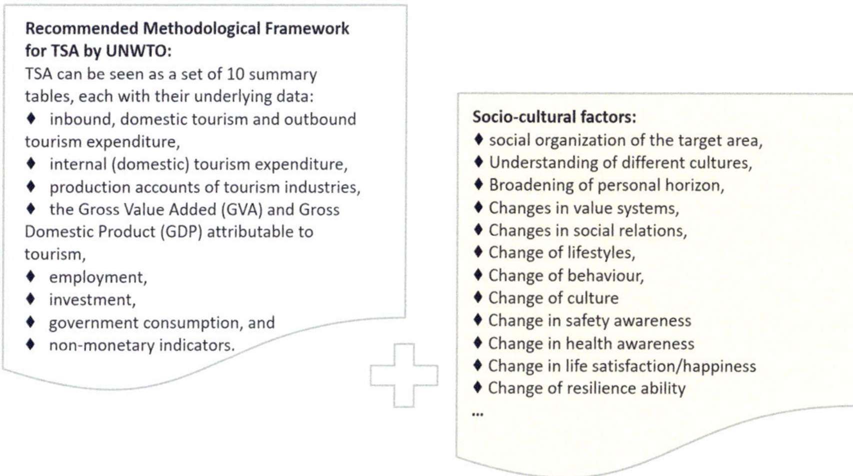

Fig. 2.1. Complementary elements for a sociocultural TSA.

Source: The authors

which points the way for a sustainable balance of work, tourism and living in the tourism region.

Against this background, it could be a challenge methodically to collect determinants of quality of life and to correlate them with tourism and economic value creation. Today, tourism satellite accounts (TSAs) represent a framework which was developed for quantifying the importance of tourism. Above all, on the basis of quantitative measures, TSAs record the range of products and services produced in tourism-relevant branches of the economy, whose goods and services are predominantly consumed by tourists – i.e. consumption and the associated economic effects. As a socioeconomic consideration of this research, it is proposed to expand TSAs or to supplement them with sociocultural effects (Fig. 2.1). The intention behind this is to evaluate and compare tourists' and natives' expectations, experiences and burdens, with the practical aim of deriving approaches for harmonizing both the socioeconomic living space and the cultural experience space of guests and locals.

On the left side of Fig. 2.1, the methodological framework of the UNWTO is presented for the purposes of recording added value in tourism. On the right side, supplementary tables are suggested to analyse and assess the perception of, acceptance of and resilience to the impacts of sociocultural tourism, which have a direct impact on quality of life. Each of these complex and subjective phenomena are addressed as complementary elements of TSAs, such as understanding of different cultures; broadening of personal horizons of both locals and guests; changes in value systems; changes of lifestyle and behaviour; and finally the change of inherited ability for resilience. These are undoubtedly very challenging fields of research. In the further consideration of a methodical recording and assessment of these qualities, corresponding measurable criteria and indices would have to be developed, which would then be correlated with the initial methodological framework for TSAs developed by UNWTO.

A sociocultural TSA could also be conducted on its own without an economic survey of the added value of tourism, especially because the methodological effort should be taken into consideration. In the well-known and large destinations affected by overtourism, a complementary approach would be appropriate. And, in smaller affected regions, a solely sociocultural TSA approach would be more feasible.

A sociocultural TSA approach would additionally support strategic and situational analysis as a crucial part of strategic planning for the development of tourist regions and habitats. Furthermore, this new approach would be able to enhance a stronger focus on the integration of the local or regional population, which is required for effective destination development. This is due to the fact that objectifiable data that is necessary for the implementation of the sociocultural TSA is actually sociocultural primary data. With its efficient flow into planning processes, the whole population can be integrated organizationally.

2.5 The Concept of Destination Resilience to Prevent Overtourism

The experts' considerations presented in the main part of this chapter lead us to an overall conclusion from the novelty of the results achieved in this study: to prevent and manage the present effects of overtourism which are perceived even acuter than ever in the situation of the COVID-19 pandemic, regional planning and development strategies should be based on the concept of *destination resilience*. In particular, as far as the negative impacts of overtourism on perceived quality of life are concerned, the dimensions of sociocultural and sociopsychological resilience of both locals and guests have to be strengthened (Hassan *et al.*, 2017). This approach would allow for one to deal with existing impairments in such a way that powerful changes in lifestyle and personal attitudes can be introduced (Traskevich and Fontanari, 2018).

The core idea of the concept of resilience for tourism destinations implies their functional autonomy and independence in the value chain of tourism products. This requires the establishment of dialectical relationships in production and consumption of tourism products, and at the same time a process of responsible, open and attentive co-living of the local community and tourists (Фонтанари and Траскевич, 2019). These dialectical interrelations include the following functional elements, which are consciously aimed at ensuring the reliability of the supply system and integrative social cooperation within the local or regional population:

- independence in the disposal of production and consumption waste, both for the tourism industry and in the everyday life of the local population;

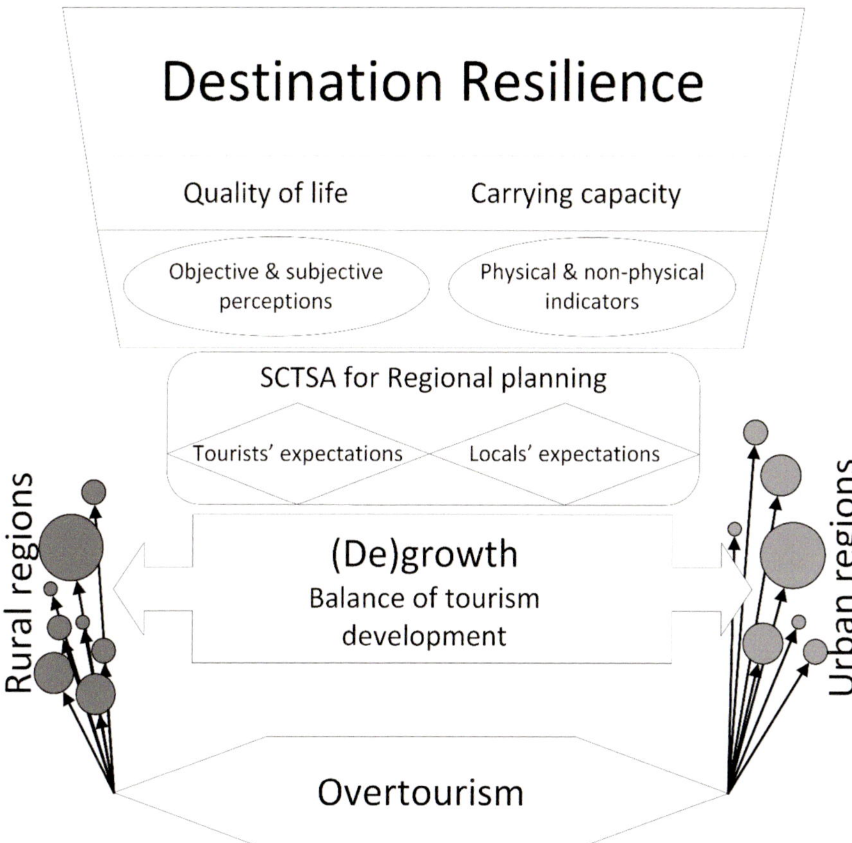

Fig. 2.2. (De)growth framework to handle overtourism: life quality and resilience approach. SCTSA, Socio-cultural Tourism Satellite Account.

Source: The authors

- autonomy of water and electricity supply;
- security of food supply;
- awareness of possible risks among the local population and tourists;
- overall personal (psychological) resilience towards any external challenges and critical situations, which are generated by systemic crises, like the present COVID-19 pandemic (Fontanari and Traskevich, 2020).

Figure 2.2 shows the framework of application of the concept of destination resilience to tackle the present effects of overtourism within strategic planning approaches that are elaborated for the development of resilient destinations (Innerhofer *et al.*, 2018; Fontanari *et al.*, 2020). In this framework, the instrumental approach of a sociocultural TSA forms the basis for the quantitative and qualitative evaluation of the prerequisites and the perspectives for building a resilient destination, with a focus on the psychological resilience of the residents and their guests.

The framework incorporates the dynamic impacts of overtourism to pointed (hot-spot) urban or rural destinations, as well as shrinking or degrowth strategies which were previously discussed and used as frequent solutions to reduce the phenomenon of overtourism. This is contrasted and merged with a development approach, which deals for the first time with the question of resilient structures (Fontanari *et al.*, 2020) and resilient orientation of a destination (Fontanari and Traskevich, 2020). The fundamental consideration of this framework – based on the empirical results of the Delphi survey – is how to make the perception of overtourism measurable, and how to correlate it with the economic value added and the positive economic effects of tourism (which above all means an increase in the prosperity of the population). The introduction of the new instrument, a sociocultural TSA, goes beyond the purely economic creation of value and deals intensively with the question of quality of life (which is a particularly topical aspect in times of pandemic) and its correlation with tourism volume and intensity. The results of such a measurement over time determine the degree of further resilient development of the destination, in order to achieve a balance between economic value creation with low leakage rates and subjective value enhancement of quality of life for both residents and tourists. In the COVID-19 pandemic crisis, this can mean more but also less tourism, implying either growth or degrowth strategies.

2.6 Conclusion

The analysis of the outcomes of the Delphi survey presents the conclusion that tourism managers and planners need to change their way of thinking and governance approaches, in order to overcome the effects of overtourism which are still perceived within the present devastating impacts of the COVID-19 pandemic. Previous findings in terms of holistic and integrative tourism planning, its methods and instruments, can still be applied in the current situation of systemic crisis and the indefinite future of the global tourism industry – under the umbrella concept of (de)growth. In addition to the economic value added, which is normally associated with trivial business growth, sociocultural value and the overall appreciation of tourism and its effects should also be taken into account.

In particular, the involvement of the local population must be based on empirical and thus meaningful data with a long-term orientation. A long-term panel on selected sociocultural added value and the attitudes of the local population will contribute to a more objective assessment of the tourism situation and development, thus having a stronger influence on strategic planning and public communication. Within this objective and comprehensive data, either growth or degrowth strategies can be implemented for both competitive and sustainable tourism development, which would above all contribute to the quality of life of the local community. This also increases the dimension of (psychological) community resilience for the destination which is especially urgent in the present times of COVID-19.

It is concluded that the coherent application of the concepts of degrowth and destination resilience contributes to solving the problem of the negative impacts of tourism. Furthermore, it is proposed to design a sociocultural TSA to measure the impact of tourism and to bring the native community into permanent self-evaluation of perceived sociocultural benefits or losses.

Furthermore, a higher destination resilience also offers new perspectives for the entire concept of destination (de)growth. The joint practical application of the concepts of degrowth and resilience bring to tourism regions the following substantial benefits:

- innovative offers in the portfolio of tourism products;
- stronger interactive communication between locals and guests; and
- an increase in the identification factor for the region itself, and thus for external marketing.

Confrontation of the ideas of (de)growth and building on resilience restores the initial identity of the destination and emphasizes its uniqueness. This is especially revealed in the development of USP tourism products on the basis of local recreational resources: these products are oriented on local identity – but not on the trendy fashionable 'touristic solutions' which are offered by globalized mass tourism. Shaping the touristic uniqueness of destinations is based on the impacts of new partnership behaviour between all participants of tourism in the region.

The conclusion obtained by the present study is that the balance between competitive entrepreneurial activity and overall happiness and health in the community is achieved for tourism destinations by means of the coherent application of both the recently discussed concepts of (de)growth and resilience. However, it should be stressed that this ideal balanced situation can only be achieved in real life if developments in infrastructure, product design and associated marketing are always seen from the perspective of different target groups and local stakeholders. Therefore, new competences are required for tourism management, so that the decision-making processes in the destination can produce positive synergies for all those who are directly or indirectly involved and affected by tourism activities. In this regard, the presented framework to handle the critical effects of tourism (or the systemic crisis of the tourism industry) clarifies the new thinking and acting which is required for destination managers. This framework is based on the instruments of well-approved participative planning and expands the perspective to tackle the negative effects of tourism by means of the concept of (de)growth. At the same time, competitiveness and safety in the implementation of this framework is assured through the orientation of regional development to produce a complementary concept of destination resilience. The instrumental key for such a harmonized resilient development is presented within a sociocultural TSA, which is introduced for an objectifiable assessment of the phenomenon of overtourism itself and the perspectives needed to overcome it.

This framework should be transferred to all the participative stakeholders: first of all, to overcome the stereotype that tourists might be seen as offenders and locals as victims. Furthermore, it can be seen as the way forward to meet the needs of opposite sides (locals and tourists) and adopt them within an 'ambidextrous management approach' (Papachroni *et al.*, 2015; Smith, 2017). Based on this framework, the strategy of regional development has to be elaborated in detail, in order to achieve a more harmonious and resilient destination that is capable of tackling and even preventing all the perverse impacts of tourism, including the systemic crisis of the tourism industry.

References

Blind, K. and Cuhls, K. (2001) Der Einfluss der Expertise auf das Antwortverhalten in Delphi-Studien: ein Hypothesentest [The influence of expertise on response behaviour in Delphi studies: a hypothesis test]. *ZUMA Nachrichten* 25(49), 57–80.

Cheema-Fox, A., La Perla, B.R., Serafeim, G. and Wang, H. (2020) *Corporate Resilience and Response During COVID-19. Working Paper 20-108*. State Street Associates; Harvard Business School.

Dodds, R. and Butler, R. (eds) (2019) *Overtourism: Issues, Realities and Solutions*. De Gruyter, Berlin, Germany.

Fontanari, M. and Berger-Risthaus, B. (2020) Problem and solution awareness in overtourism: a Delphi study. In: Pechlaner, H., Innerhofer, E. and Erschbamer, G. (eds) *Overtourism: Tourism Management and Solutions*. Routledge (Taylor & Francis Group), Oxford, UK, pp. 43–67.

Fontanari, M. and Traskevich, A. (2020) *The Concept of Destination Resilience: Application for advanced sustainable Management in Tourism*, Manuscript submitted for publication. International School of Management, Cologne, Germany; Hochschule Fresenius University of Applied Science, Cologne, Germany.

Fontanari, M., Traskevich, A. and Kutsch, H. (2020) *Corporate resilience within tourism enterprises*, Manuscript submitted for publication. International School of Management, Cologne, Germany; Hochschule Fresenius University of Applied Science, Cologne, Germany.

Gössling, S., Scott, D. and Hall, C.M. (2020) Pandemics, tourism and global change: a rapid assessment of COVID-19. *Journal of Sustainable Tourism* 12(5), 1–20. DOI: 10.1080/09669582.2020.1758708.

Gretzel, U. (2019) The role of social media in creating and addressing overtourism. In: Dodds, R. and Butler, R.W. (eds) *Overtourism: Issues, Realities and Solutions*. De Gruyter, Berlin, Germany, pp. 62–75.

Haifeng, Y., Jing, L. and Mu, Z. (2012) Rural community participation in scenic spot. A case study of Danxia mountain of Guangdong, China. *Journal of Hospitality & Tourism* 10(1), 76–112.

Hall, C.M., Scott, D. and Gössling, S. (2020) Pandemics, transformations and tourism: be careful what you wish for. *Tourism Geographies* 22(3), 577–598. DOI: 10.1080/14616688.2020.1759131.

Hammond, M. and Wellington, J. (2013) *Research Methods. The Key Concepts*. Routledge (Taylor & Francis Group), Oxford, UK.

Hassan, A., Ramkissoon, H. and Shabnam, S. (2017) Community resilience of the Sundarbans: restoring tourism after oil spillage. *Journal of Hospitality and Tourism* 15(1), 1–18.

Innerhofer, E., Fontanari, M. and Pechlaner, H. (eds) (2018) *Destination Resilience: Challenges and Opportunities for Destination Management and Governance*. Routledge (Taylor & Francis Group), Oxford, UK.

Innerhofer, E., Erschbamer, G. and Pechlaner, H. (2020) Overtourism: the challenge of managing the limits. In: Pechlaner, H., Innerhofer, E. and Erschbamer, G. (eds) *Overtourism:*

Tourism Management and Solutions. Routledge (Taylor & Francis Group), Oxford, UK, pp. 3–14.

Jacobsen, J.K.S., Iversen, N.M. and Hem, L.E. (2019) Hotspot crowding and over-tourism: antecedents of destination attractiveness. *Annals of Tourism Research* 76, 53–66. DOI: 10.1016/j.annals.2019.02.011.

Jin, X.C. and Sparks, B. (2017) Barriers to offering special interest tour products to the Chinese outbound group market. *Tourism Management* 59, 205–215.

Koens, K., Postma, A. and Papp, B. (2018) Is overtourism overused? Understanding the impact of tourism in a City context. *Sustainability* 10(12), 4384.

Lapointe, D. (2020) Reconnecting tourism after COVID-19: the paradox of alterity in tourism areas. *Tourism Geographies* 22(3), 633–638. DOI: 10.1080/14616688.2020.1762115.

Milano, C. (2017) *Overtourism and Tourismphobia; Global trends and local context*. Ostelea School of Tourism & Hospitality, Barcelona, Spain.

Moscardo, G., Konovalov, E., Murphy, L., McGehee, N.G. and Schurmann, A. (2017) Linking tourism to social capital in destination communities. *Journal of Destination Marketing & Management* 6(4), 286–295.

Muler Gonzalez, V., Coromina, L. and Galí, N. (2018) Overtourism: residents' perceptions of tourism impact as an indicator of resident social carrying capacity – case study of a Spanish heritage town. *Tourism Review* 73(3), 277–296. DOI: 10.1108/TR-08-2017-0138.

Nepal, S.K. (2020) Adventure travel and tourism after COVID-19 – business as usual or opportunity to reset? *Tourism Geographies* 22(3), 646–650. DOI: 10.1080/14616688.2020.1760926.

Niederberger, M. and Renn, O. (2018) *Das Gruppendelphi-Verfahren [The group Delphi method]*. Springer Verlag, Wiesbaden, Germany.

Niewiadomski, P. (2020) COVID-19: from temporary de-globalisation to a re-discovery of tourism? *Tourism Geographies* 22(3), 651–656. DOI: 10.1080/14616688.2020.1757749.

Oklevik, O., Gossling, S., Hall, C.M., Kristian, S.J.J., Petter, G.I. *et al.* (2019) Overtourism, optimisation, and destination performance indicators: a case study of activities in Fjord Norway. *Journal of Sustainable Tourism* 27(12), 1804–1824.

Panayiotopoulos, A. and Pisano, C. (2019) Overtourism dystopias and socialist utopias: towards an urban armature for Dubrovnik. *Tourism Planning & Development* 16(4), 393–410.

Papachroni, A., Heracleous, L. and Paroutis, S. (2015) Organizational ambidexterity through the lens of paradox theory: building a novel research agenda. *The Journal of Applied Behavioral Science* 51(1), 71–93.

Park, S.-Y. and Petrick, J.F. (2006) Destinations' perspectives of branding. *Annals of Tourism Research* 33(1), 262–265. DOI: 10.1016/j.annals.2005.10.007.

Smith, S.M. (2017) Organisational ambidexterity: welcome to paradox city. *Human Resource Management International Digest* 25(1), 1–3.

Traskevich, A. and Fontanari, M. (2018) Mental wellness in resilient destinations. *International Journal of Spa and Wellness Industry* 1(3), 193–217.

Wall, G. (2019) Perspectives on the environment and overtourism. In: Dodds, R. and Butler, R.W. (eds) *Overtourism: Issues, Realities and Solutions*. De Gruyter, Berlin, Germany, pp. 27–45.

Zukunftsinstitut (2020) *Die Welt nach Corona. Business, Märkte, Lebenswelten – was sich ändern wird* [The world after Corona. Business, markets, lifestyles – what will change]. Available at: https://onlineshop.zukunftsinstitut.de/shop/die-welt-nach-corona/

Фонтанари, M. and Траскевич, А.Г. (2019) Концепция резильентности дестинаций и перспективы ее применения в отношении сельских дестинаций Беларуси. *Белорусский экономический журнал* 1, 136–151[Fontanari, M. and Traskevich, A. (2019) The concept of destination resilience and prospects of its application with regard to rural destinations in Belarus. Belorusskij Ekonomicheskij Zhurnal = Belarusian Economic Journal, 1, 136-151. (in Russ.)].

3 Community-based Tourism in a Degrowth Perspective

Ernest Cañada[1]*

Alba Sud - Investigation and Communication for Development, Barcelona, Spain

3.1 Introduction[2]

In recent years the debate on tourism degrowth has increased its presence substantially in the agenda of social movements, in the media and in the academic field. The discontent and discomfort with the performance and the results of tourism development have provided a greater audience for transformation demands of these activities. The discussion has been focused on urban destinations which have suffered the effects of tourism overcrowding. In this chapter, however, the view is broadened towards rural territories. The vulnerability that some of these territories have shown, with strong decreases in tourism activity due to external phenomena, such as the case of the COVID-19 pandemic since early 2020, has manifested the need to build more resilient rural tourism proposals. In this context, the debate on tourism degrowth in rural regions acquires a specific sense.

Through the analysis of three community-based tourism experiences in Central America – Cooperativa Los Pinos (El Salvador), Ecoposada El Tisey (Nicaragua) and Stribrawpa (Costa Rica) – the potentialities and limitations of a degrowth perspective in the rural context are analysed. The three selected cases share recognition in the business success of the offer, even if distant from more traditional administration models. The analysis and discussion of these cases could enrich the debate on tourism degrowth and, at the same time, the analysis of community-based tourism from other perspectives.

3.2 Literature Review

The touristic pressure in some urban destinations, accentuated by the transformations of capitalism due to the global financial crises that began back in

*ernest@albasud.org

DOI: 10.1079/9781789245073.0003

2008, have sharpened manifestations of social discomfort with this activity. The growing discontent and the social mobilization that appeared in some territories have facilitated critical reflection in favour of the transformation of territories with a strong presence of tourism and its activity (Milano *et al.*, 2019). In some cases, this has meant an increase in the concern to sensitize all the actors in the tourist chain towards more responsible forms of tourism (Burrai *et al.*, 2019) or to reduce the dynamics of exclusion in favour of inclusive tourism (Scheyvens and Biddulph, 2018). However, in a progressive way the paradigm of 'tourism degrowth' has also acquired relevance. This concept arises from a theoretical approach, degrowth, which questions the logic of unlimited growth and points to the need to boost a process of political and social transformation which reduces the use of materials, energy, technological resources and work, as well as the improvement of the quality of life of the population and more efficient use of territories (Latouche, 2006, Latouche, 2007). Degrowth has had multiple interpretations and influences from other schools of thought (Andriotis, 2018). Despite this, enough consensus has been reached to understand that it fundamentally implies decolonializing imaginaries vastly settled in the paradigm of growth (Demaria and Bakker, 2019) and voluntarily transitioning towards another model of social organization. From this perspective, processes of desired transformation vs the risk of cyclic capitalism crises imposing other forms of forced degrowth are advocated (Blanco-Romero, 2019).

In the field of tourism, the logic of unlimited growth has been predominant, with a large consensus up until recently, despite critical voices (although a minority have questioned it since the 1970s) (Cañada and Murray, 2019). In contraposition, the paradigm of tourism degrowth implies the reduction of tourism's weight and the reorganization of societies highly dependent on this type of activity, with justice and equity criteria that, at the same time, allow the shielding of tourist experiences of the best quality (Andriotis, 2014, 2018; Gössling and Peeters, 2015; Fletcher *et al.*, 2019; Higgins-Desbiolles *et al.*, 2019). The search for new forms of tourism degrowth, more equitable and inclusive, is considered essential to face the danger that degrowth is simply achieved through the rise of prices and the gentrification of tourism (Fletcher *et al.*, 2019).

In practical terms, tourism degrowth proposals have been translated into two huge lines of action: containment of growth and proposals to organize tourism production in other ways.

On the one hand, the demands of tourism growth contention mechanisms, such as the regulation of accommodation capacity, the limitation of transport infrastructures or an increase in taxation, mainly in accommodation and mobility (Fletcher *et al.*, 2019), have been reflected in some tourism policy initiatives, in addition to being part of the social movements' demands agenda in cities in the south of Europe (Gómez and Pardo, 2019). In a more moderate sense, this orientation has also been expressed in some public policies, as touristic territorial planning. In certain territories of the Spanish Mediterranean, both on the coast and in urban destinations, there has been an attempt, even if with limited results, to apply more restrictive criteria to tourism development (Blázquez-Salom *et al.*, 2019).

On the other hand, it also means a change in methods of travel and the organization of tourism activity. The starting point of such change is based on the interest of some tourists in taking part in other practices, more oriented to quality rather to quantity, which reject traditional forms of tourism and result in fewer resources and service consumption, so that territories are able to develop themselves without depending on tourism and its constant growth (Andriotis, 2018). Partially, this has been favoured by the new forms of tourism organization and consumption under a post-Fordist scheme (Ioannides and Debbage, 1997, 1998; Egresi, 2016). By itself, post-Fordism does not lead to this change of mentality, but it has allowed for the revalorization of other ways of travelling and tourism interest for singular and differentiated experiences. This way, minority forms of organization of tourism activity are recognized and new formulas are tested, such as slow tourism, which recommends reducing the rhythm of tourism consumption, spending more time in a destination and integrating in a friendlier way (Clancy, 2017). This perspective is more considered as an attitude when travelling rather than a typology of tourism or tourist (Oh *et al.*, 2016; Özdemir and Çelebi, 2018; Serdane *et al.*, 2020). It also implies a higher assessment of the quality of the experience instead of the quantity (Timms and Conway, 2012). Staycation has also been promoted, understood as a way of organizing a holiday from home instead of going on an overseas trip. This also includes leisure activities and daily excursions to close-by destinations, but spending the night in the usual residence (Moltz, 2009; de Bloom *et al.*, 2017; Heimtun, 2017). This activity could create more benefit than travelling somewhere else (Andriotis, 2018) and boost the economy at a local level (Izcara and Cañada, 2020). In this scheme of revalorization of certain tourism practices, culinary tourism is also found, the main interest of which revolves around food culture and the production of food products. Moreover, this type of tourism may act as a stimulus for local production (Gascón, 2018).

The analysis of new forms of tourism production and consumption could be a fertile line of investigation, oriented by the need to build real utopias to capitalism, paraphrasing Erik Olin Wright. This would entail the development of an emancipatory social science applied to the analysis of tourism, the role of which would be to generate knowledge to put into question oppression, dispossession and exclusion, and, at the same time, favour processes that would eliminate them and allow for the development of human potentialities through social action (Wright, 2010).

Located in this framework, in rural areas, there have been expressions of discomfort and conflict due to tourism development, which has impacted the agenda of social movements (Milano, 2018; Milano *et al.*, 2019). The literature concerning dispossession and socio-ecological processes in rural areas linked to tourist development is broad, with detailed examples both in the south of Europe from the 1970s (Kousis, 2000), as well as later in different locations of Latin America, to quote just the geographic area of the present research, such as Central America (Cañada, 2018), Peru (González, 2018), Colombia (Ojeda, 2016) and Mexico (Marín, 2015). In this tradition, more or less explicitly, there is a claim for the necessity to reduce touristic pressure and the processes of dispossession of livelihood components which have come with it, connecting with the thoughts of tourism transformation and even with degrowth.

However, the contribution that diverse experiences in a rural context that have opted for non-conventional practices and strategies could have to tourism degrowth has been less explored. Many of these initiatives take place in low touristified places which, because of their socio-territorial resilience – understood as the capacity of equilibrating through prevention and adaptation (Blanco *et al.*, 2019) – have been able to innovate and transform traditional forms of organizing tourism activity. In this sense, small-format initiatives stand out, such as community-based tourism which, due to its dimension and variety, can bring innovative ideas that help to rethink a transition in the forms of organizing tourism activity.

Regarding community-based tourism – understood as a form of management in which rural communities exercise control in the decision making on how the activity is organized and how its benefits are redistributed (Cañada, 2014; Ruiz-Ballesteros, 2017) – this has been subject to a strong debate regarding its viability and the implications of its development (Cañada and Murray, 2019). One of the aspects that has not been addressed, despite the large body of existing literature, is the contribution of community-based tourism in a degrowth perspective, and under which conditions this could happen. In one of the few texts that has highlighted this relation, it is outlined that community-based tourism helps maximize the benefits of tourism activity, either for organized local villages or for tourists themselves. It is also identified as a way of organizing tourism activity that boosts a perspective of tourism degrowth (Andriotis, 2018). Two recent empirical studies on community-based tourism in Ecuador, one in the community Kichwa Añangu (Renkert, 2019) and the other in the canton of Cotacachi (Chassagne and Everingham, 2019), link this management model with the postulates of degrowth. In both cases it is emphasized how community-based tourism is integrated in a worldview different from the traditional capitalist scope. Under schemes such as the 'good life', community-based tourism appears to be part of a collective strategy to solve the needs of the population complemented by other activities defined by systems of collective governance.

Despite this, community-based tourism also presents some limitations from an emancipatory perspective. Two critical elements emerge in the debate about this type of experience. First is its main orientation towards affluent tourists, who make it difficult for middle- and lower-class tourists from their own countries to access this type of offer. Second is how the targeting of this type of client strengthens their dependence on international long-distance travel. The growing trend in the tourism market, oriented towards post-Fordist forms, has increased the marketing opportunities for community-based tourism, and therefore the interest of tour operators and wholesale agencies. Likewise, the predominant orientations of international cooperation agencies and public policies have reinforced this emphasis towards an international affluent clientele as the best option for community initiatives. As a result, these social and climatic reasons coalesce in a critique of the emancipatory potential of community-based tourism (Cañada, 2019a).

However, not all community-based tourism initiatives are geared toward this type of clientele. In Central America, there are commercially successful endeavours that have been developed with a preference for middle- or lower-income sectors, and even those initiatives aimed at foreign markets with greater purchasing power have incorporated redistributive and degrowth measures with high emancipatory potential. In some cases, this option has been selected due to a need for adaptation in a context that did not allow for other options, but has also been a product of collective thought looking for non-conventional strategies.

3.3 Objectives and Methodology

The main goal of this chapter is to explore the potentialities and limitations of community-based tourism in Central America from the perspective of tourism degrowth. The proposal is to identify non-conventional forms of tourism development that might help us think about community-based tourism from new perspectives, more in line with global concerns about tourism degrowth.

I analysed three community-based tourism initiatives: Cooperativa Los Pinos in El Salvador, Ecoposada El Tisey in Nicaragua and Stribrawpa in Costa Rica. The analysis was developed in the framework of a wider investigation, not adjusted to its possible relation to degrowth. In this investigation, nine community-based tourism initiatives were studied for over ten years. Apart from the above mentioned, there were also: Bosque de Cinquera and Ruta de Paz in El Salvador, UCA Miraflor and Finca Magdalena in Nicaragua and Los Campesinos and ASOPROLA in Costa Rica. These initiatives were selected following these criteria: being operative in the tourism business for more than five years; regularly receiving tourism income and no dependence on cooperation projects (although they might have received diverse external support); and diversity in the forms of organization. It was recognized that these initiatives needed to have succeeded in their main goal: developing a regular commercial activity that might have allowed them to generate and diversify their income. Other forms were rejected, such as those where a commercial failure had taken place and tourist activity had stopped. The option was to analyse the social effects derived from the implementation of new economic activity with a reasonable degree of consolidation. From this investigation, it is possible to present case studies (Cañada, 2017a), a comparative analysis of the implications in terms of gender relations (Cañada, 2019b) and the most prominent practices (Cañada, 2014), while other works are in progress.

As noted, for this study on community-based tourism and degrowth, three initiatives, one in each country, were selected. The reason is that they have taken measures that can better contribute to this debate. In each of these selected initiatives, successive fieldwork visits took place, with a duration of one or two weeks in each, between 2009 and 2019 (Table 3.1).

During the fieldwork visits, interviews with different members of the initiatives were carried out, and they were repeated in most cases in successive

Table 3.1. Fieldwork visits by initiative.

Initiative	Visits	Periods of investigation
Cooperativa Los Pinos (El Salvador)	4	May 2009 February 2014 June 2015 September 2018
Ecoposada El Tisey (Nicaragua)	4	March 2009 March 2014 June 2015 December 2018
Stribrawpa (Costa Rica)	3	February 2014 May 2015 August 2018

Source: own elaboration.

encounters (Table 3.2). Basically, the most involved and active persons in each initiative were interviewed, either those with labour links, but also bearing in mind the need to contrast their vision with the people that had minor responsibility roles or who were not linked to the tourism activity, or even to the community experience. The variation in the number of interviews in each initiative is due to the differences in the numbers of individuals involved. In the case of Ecoposada El Tisey, the reduced number of people interviewed, compared to other initiatives, is because this is a family-based initiative and therefore there are fewer members. In the three initiatives, the first fieldwork visits were focused on the history of each organization. The causes of any changes brought about in the productive and organizational areas, as well as the evolution of this over time, were attempted to be identified. A special focus was put on the internal organization of both the global initiative and in relation to the management of tourism activity. In the same way, results were gradually obtained resulting from decisions to develop tourism activity. With every person interviewed, apart from these general aspects, the individual role of each was identified, as well as their vision and thoughts regarding every topic tackled. Moreover, the interviews were used to contrast pieces of information supplied by every member, and these were triangulated with what was obtained from other sources. In the following visits the interviews, which mainly focused on the same members, were used to update the information and identify possible changes in interviewees' respective visions.

Table 3.2. Interviewees differentiated by gender and initiative.

Initiative	Total	Male	Female
Cooperativa Los Pinos	14	9	5
Ecoposada El Tisey	4	2	2
Stribrawpa	23	13	10

Source: own elaboration.

Furthermore, as time went by, a relationship based on trust emerged, which allowed for greater accuracy in the analysis.

Apart from the interviews, other sources of information were used. In each of the cases a short-term ethnographic work was developed, which was reflected in a field diary, where observations and informal conversations were recorded. In the majority of the initiatives, there was also access to informal documents (constitutional documents, minute books, visitors registers) and in the case of Cooperativa Los Pinos, the full business accounts were consulted. In all cases, there was at least one mixed group discussion with the majority of the members of the initiative on the functioning of tourism activity and the diverse impacts it had had. These discussion groups were useful in presenting some of the main conclusions for collective analysis.

In the present text, besides describing the selected initiatives, the identification of adopted strategies that can be considered examples for a debate on how tourism can be rethought in a degrowth and post-capitalist perspective has been prioritized.

3.4 Results

3.4.1 Case of Cooperativa Los Pinos (El Salvador)

Cooperativa Los Pinos is located in the municipality of El Congo, in the department of Santa Ana, in El Salvador, by Lake Coatepeque. It was constituted in 1980 and nowadays is composed of 87 members – 70 men and 17 women. It has an extent of 285 hectares, primarily dedicated to the production of coffee. The main income of the cooperative comes from selling of coffee and tourism services (Cañada, 2017a).

Before being a cooperative, the area was the property of Rafael Guirola, a member of one of the most powerful 14 families in El Salvador. This coffee oligarchy had run the country for decades and its power was based on a profound inequality in access to land. The situation of the former workers was extremely poor, without access to tap water or electricity, and housing was very inadequate.

> Here there was a settlers' system. The families that lived in the fields of the farmlands did not have the guarantee of a permanent job. Some days one worked and, suddenly, works stopped with no explanation. They just said there was no job, and that everybody should find their own solution. In the houses, there was no access to electricity, only in the house of the boss. There was no access to water either. Each family had a pitcher, about 30 bottles of water, in the morning, and the same in the afternoon. This was all the access for a family, and it was not enough even to cook. The majority of the people had to go to the lake to clean their clothes, to wash. And naturally there was no access to health either. It was a pretty precarious situation. Now many are unaware of the real situation lived at that time. (SG02, partner and technician of the Cooperativa Los Pinos, personal communication)

The existing inequality and the miserable conditions in which the majority of the rural population lived created an environment of social agitation, frequently suppressed by different governments and the army in favour of the interests of the oligarchy groups. In 1979 there was a putsch in El Salvador where the Junta Revolucionaria de Gobierno (Revolutionary Governing Board) seized power, intending to implement social reforms that might help to prevent the advance of the popular organization, the guerrilla movement Frente Farabundo Martí para la Liberación Nacional (FMLN) (Farabundo Martí National Liberation Front).

> Nobody knew anything of the agrarian reform, or what its objective was. This was sort of a smokescreen to stop the rebellion of the people. (TH01, partner of the Cooperativa Los Pinos, personal communication)

A short time after this, the 5 March 1980, a Basic Law of the Agrarian Reform was enacted. It established that nobody could own more than 100 hectares, leading to the redistribution of land to tenant farmers. Cooperativa Los Pinos was created in this context, with tenants who worked on the farm at that time. With this measure, the state aimed for the deactivation of the potential bases of the insurgent movement. However, without real support to the cooperatives recently created, failure of the initiative seemed likely and a progressive return of the land to the original owners was foreseeable.

Despite this, Cooperativa Los Pinos managed to survive and carry out diverse social projects which countered the instability and precarity the tenant farmers used to live in. The construction of accommodation, having soil to cultivate, the arrival of electric power, the possibility of going to a health centre in emergency situations or the support of education are some of the improvements that were especially highlighted by members.

> When the cooperative opened the opportunities in education did too. Many of us have been beneficiaries of decisions that were made by the management boards and the assemblies at that time. They hired teachers so that they came to teach classes in the cooperative. They also gave the transport to go study outside the cooperative, when it was necessary. And that allowed for many of the partners to be educated and to create different conditions, with a more corporate vision. (SG02, partner and technician in the Cooperativa Los Pinos, personal communication)

From its origins, the cooperative worked in the production of coffee, giving continuity to the specialized production of the expropriated farm. In 1992, the cooperative earned a benefit which helped it to save costs and sell in better conditions, in addition to being able to process the coffee of other producers. Despite these signs of progress, the fall of international prices of coffee in 2001 from 150 the previous year to 40 dollars per quintal on the New York Stock Exchange (Flores *et al.*, 2002) heavily affected them. This provoked the decision to go a step further and extend and diversify their productive sources. On the one hand, trying to guarantee more stability in the selling process through the market of fair trade and, on the other hand, introducing touristic services.

From that moment on, they started to sell the larger part of their coffee through the market of fair trade in Europe. One of the virtues of this market is that, apart from guaranteeing more stability in the process, it provides a premium for the volume sold which must be invested in social benefits both for the members of the cooperative and for the hired workers, mainly from nearby communities. However, after the harvest of 2013–2014, production strongly diminished due to a plague of mildew, which affected many plantations in Central America. This fungus had an impact on the volume produced and the quantity of coffee that could be sold to the fair trade market. The coffee damaged by borer beetle, broken or bitten by animals, could not be sold to this market. The consequence was a significant reduction in global incomes generated by coffee in the cooperative.

After the international price of coffee fell in 2001, they tried to start up touristic services. However, this option was not fast to develop. When the cooperative put touristic activities into operation after 2007, they decided that the management had to be collective, as with the other activities on the farm.

> Since our beginning, we have been a collective cooperative and the effort has always been common good. And with tourism, we considered it was way more feasible to act the same way because then we already had experiences that had improved the business capacity of the cooperative. We work the farm collectively and the benefit is also collective, so tourism is the same. (SG02, partner of the Cooperativa Los Pinos, personal communication)

In order to kick off tourism in the cooperative, they built a kitchen and a restaurant with a view of Lake Coatepeque, some cottages, a path from the restaurant to the lake with viewpoints and a pier equipped with three motorboats. They also built a communal hall for organized events. They bought the necessary equipment for the kitchen, the restaurant and the cottages. All these infrastructures were financed with grants from PREMODER, a programme run by the Ministry of Agriculture, the social benefit premium from the coffee sold through fair trade, and funds from the cooperative. The first buildings went up between 2008 and 2009, and in the first week of September 2009 they finally opened to the public. As tourism started to generate income, they could improve the infrastructures and better maintain them by fixing and fencing the paths.

The cooperative has a strategic plan in which they have displayed what they need and which investments they should execute to consolidate the tourist offer. Furthermore, to manage tourism, the management board of the cooperative (Consejo de Administración) created a tourism committee. It meets once a month and is used to facilitate coordination and generate the information needed by the board to make decisions. This organ is composed of a president, who is part of the board, a chef, a cashier and one member of the vigilance committee.

Tourists who visit Los Pinos are mainly Salvadorans who live in San Salvador and who can reach the farm in about 45 minutes in a private vehicle. They also receive visits from Salvadorans who live abroad, especially in the USA, and who go home for their holidays, as well as some foreigners. It is remarkable that clientele linked to different public and private institutions (universities, schools, churches) have found in the Cooperativa Los Pinos an adequate place to organize activities. In the main, they are middle class. Since the beginning, managers of the cooperative clearly understood that this was the profile of clients they were looking for.

> We saw that our target was middle class, not high middle nor low middle, simply middle class. We wanted to be a place where those who cannot go to expensive places could go. If someone is below middle class they cannot afford a restaurant, or if they can they do not because their priorities are different. (OM02, partner from the Cooperativa Los Pinos, personal communication)

The numbers of visitors and tourists to the cooperative have positively evolved, and in 2019 they exceeded 20,000 people. These visitors initially became aware of the place through recommendations of other people who had already visited and through documentaries broadcast on local media. Lately, the promotion of social networks has incremented these visits. They are usually restaurant clients who, in some cases, take part in other activities. In this sense, half-day visits are the base of the cooperative's touristic business. This positive evolution is reflected in the generated income from tourism. It is both highlighted in the progressive increment in the total income and benefits generated, as well as the stability of this progression. Since 2014, more than 80% of tourist income has been generated by the restaurant. The positive evolution of tourism income, which in 2018 reached close to US$200,000, compared to the approximate $300,000 coming from coffee, has delivered another relevant source of income, and helped to reduce the impact of fluctuations in the price of coffee that, due to diverse causes, are very high.

Thanks to tourism, the production of coffee in the cooperative has found new markets, from which they receive even more than from fair trade commerce, as selling toasted powdered coffee bags or serving cups in the restaurant is more profitable.

> In the last two years, the selling of toasted powdered coffee has increased. Demand is stronger either by people who come from the outside and from within the community who order some. Before tourism, we did not have this amplitude. Therefore, tourism has also come to position the part of toasted powdered coffee in the cooperative. (JRR01, partner of the Cooperative Los Pinos, personal communication)

Due to the great affluence of visitors, the restaurant constantly needs to provide numerous foodstuffs, and this has stimulated new productive activities. Therefore, the cooperative has started up a greenhouse to produce vegetables and a pond for tilapias (local fish). Other products are bought, when possible, in the community. One of the problems they have encountered in broadening their purchasing from local communities is the difficulty of assuring continuity of supply and quality.

The partners of the cooperative are satisfied with the introduction of tourism and the changes it has brought. They emphasize the fact that new jobs and opportunities for members of the community have been created and that the dependency towards coffee has been reduced.

Location factors of the touristic activity (close to a natural and frequented attraction and proximity to good communication links with a huge urban centre containing potential tourists) have been important for success. These are favourable conditions, but not determinant. On this basis, a strong collective organization process, business vision, capacity and investment of own resources and connections to a potential market are funded.

Thanks to the cooperative, the life of the people and their families has significantly improved. In recent years, another measure has been implemented so that the elderly who cannot work receive a pension from cooperative funds.

> If there had been no cooperativa, we would probably not live in this place now. We would have had to leave ... somewhere else. Who knows? (LNG, partner of the Cooperativa Los Pinos, personal communication)

3.4.2 Case of Ecoposada El Tisey (Nicaragua)

Ecoposada El Tisey is a family farm located 25 kilometres from the city Estelí, in the protected area El Tisey-Estanzuela. It covers 56 hectares. There, organic agriculture, research and biotechnologies development, ranching and flower production are combined with a tourist offer that has been the main source of income in recent years. Ecoposada is frequented by Nicaraguans, especially during weekends when a lot more visitors arrive to have lunch and walk around the farm. Half of the land is dedicated to forest conservation, making it even more attractive for visitors. Ecoposada El Tisey also welcomes numerous groups who use the installations for educational activities, mainly on the theme of organic agriculture.

The farm is the property of nine brothers of the family Cerrato Jirón, who inherited it from their father at the end of the 1990s. At that time, the land was dedicated to harvesting camomile, basic seeds, coffee, tomato and cabbage. Previously, during the 1980s, a harvest of potatoes was introduced in the area, accompanied by environmental destruction due to deforestation and the use of agrochemicals. However, disease diminished production in any case and, in the end, the venture ceased to be profitable and was wound up.

When they inherited the farm from their father, the Cerrato brothers decided to maintain unity and not divide it into individual smallholdings for each of them.

> When we received it, we did not want to do plots and divide it among the heirs that were alive. Dividing it into little pieces meant hurting the forest and changing the use of the soil. (NC02, co-owner of the Ecoposada El Tisey, personal communication)

In 1992, the area was declared to be a *reserva natural* (natural reserve) (El Tisey-La Estanzuela) and the farm, being inside the margins, was obliged to change its traditional production practices. Therefore, they reforested areas that had been lost due to the harvesting of potatoes. Another factor favouring the ecologic recuperation was that in the middle of the 1990s an institution chose to develop soil conservation practices and experiment with different technologies for organic agriculture.

Touristic activity began in 2001, via the initiative of a foundation in charge of the management of the natural reserve, which wanted to promote environmentally friendlier sources of income among the population.

> This was induced mainly by the organization that co-managed the *reserva*, FIDER, Fundación para el Desarrollo Rural (Foundation for the Rural Development). They had a component that the USAID had donated, I think it was 20,000 dollars. With this component, FIDER tried to diversify the reserve because changes needed to be made, to do other activities that were not destroying the environment, sowing, polluting, creating more vegetable gardens, then the component tourism enters. They had this money, but to loan, for credit, so we were proposed to work, because we had the attraction of the viewpoint, and like this, half begged, some wanted to some others did not, we began to work. (NC02, co-owner of the Ecoposada El Tisey, personal communication)

The first visits were groups brought by different institutions and NGOs to develop capacitation processes. Members of the initiative admitted that when first receiving visitors they felt very insecure.

> In the beginning, it was not clear that this was going to be a source of income because we did not know exactly what tourism was, and we could not identify the tourist, we thought the tourist was the *chele*, the foreigner, we did not see the local or national tourist, we did not know their demands or how to attend them ... That is why we started with something small, a simple construction of bricks and chainsawed wood that I built myself. As we kept identifying the demands of the people that visited us and generating income, we started to think of other kinds of conditions. It has been a process where the investments have been done slowly, because we could not do a lot if the demand was not going to pay for it. (DC02, co-owner of the Ecoposada El Tisey, personal communication)

Over the years, in addition to collective accommodation, they built cottages with more amenities. Nowadays, they have 45 beds distributed in two communal halls and nine cottages available. All these structures were built from their own resources or through small microfinancier loans, some government programmes and individual funding from some of their members, but they did not receive international donations or government funding. The restaurant has been very successful in attracting visitors, especially at weekends, when it can serve over 100 people. The majority of visitors are Nicaraguans who have heard of the farm through word of mouth, media presence and NGO promotion. In lower numbers there are also foreign visitors, mainly young backpackers and, through some tour operators, elderly people. It is important to mention that they also receive some people engaged in scientific tourism activities, mostly interested in agriculture technologies.

The number of visitors was stabilized at around 10,000 people per year, while accommodated visitors could vary between 1200 and 1500. However, due to the political crisis in April 2018, with over 350 murders, thousands exiled and strong state repression, visitor numbers diminished. Despite this, the farm remained open and managed to maintain the business. After some months, they had 80% of visitor numbers compared to the previous year. Contrarily, community-based tourism initiatives in Nicaragua oriented to international tourism mainly had to close, at least partially.

Labour on the farm is organized among family members according to their abilities and interests. The most important decisions are made by all of them through consensus. Before the political crisis of 2018, Ecoposada was staffed by six family members and between four and six people on permanent hire to serve in the restaurant and kitchen, and perform cleaning and field tasks. In high season, temporary workers were hired (other members of the family or neighbours in the community). Salaries took as a reference those regulated by law – around 100 córdobas per day, to which they added three meals and accommodation, since most of them stayed on the farm. There were also some perks, such as vacations or bonuses, among others, although they do not have insurance.

This strong presence of visitors in the restaurant has stimulated agricultural production on the farm, especially corn, coffee, vegetables and fruits (such as oranges, lemons, bitter oranges and bananas).

> With the corn, we sow creole varieties to harvest a quality product, with which we produce *rosquillas* and other products. Just with *rosquillas,* we occupy 20 quintals per year, but if we include the cost in *tortillas, atol* and *elotes*, the total consumption of the business can reach 55 quintals … , although not everything produced is consumed in the farm. (DC02, co-owner of the Ecoposada El Tisey, personal communication)

Direct sales have been boosted by visitors. The farm sells products from the vegetable garden such as carrots, beetroot, broccoli, lettuce, coriander, parsley, spring onions and courgettes. They also sell milk, junket, flowers, ornamental plants and orchids. Coffee is commercialized under their own brand, Café Tisey. Visitor demand for food has also stimulated local supply.

> I daily sell to the Ecoposada from 10 to 15 pounds of curd cheese and sometimes eggs. Before, every eight days, I had to send the production to Estelí by bus. Now I sell it right here and I save up the ticket. (BC01, producer, neighbour of the Ecoposada El Tisey, personal communication)

In organizational terms, members say that their model, based on a family structure inserted in the community, is more efficient than other forms of collective organization.

> The family structure is better because it is the one that works for its self-interest and, in the case of cooperatives and associations, one does not feel like the owner and, for this to development you need to feel so, otherwise,

> it does not work. We are inserted in a community not just because we are members of it, but also because we are community activists. We always participate in the different activities planned to develop it. Moreover, we buy what is produced in the community. And the community does not have to pay a ticket to sell it cheaper, which is also part of the agreement we have here. (DC02, co-owner of the Ecoposada El Tisey, personal communication)

One of the characteristics of the family who has developed the Ecoposada El Tisey is the high academic and cultural level of some of its members. Some of them studied in the university and worked as professionals before starting up the tourism business and, in fact, they did not settle on the farm until they began this new activity. One of the brothers has become an expert in the investigation of technologies linked to organic agriculture, and has collaborated with different universities in this respect.

3.4.3 Case of Stribrawpa (Costa Rica)

Stribrawpa is an association and community-based tourism initiative led by women of the Bribri community, although men participate too. It is located by the River Yorkín in Costa Rica, very close to Panama. The community is also called Yorkín and is formed from little more than 200 people. The Bribri community is organized in family clans and it is the women who pass on the link to the clan.

In the 1970s, the Bribri community in Yorkín was affected by a fierce plague of the fungus *monilinia*, which ruined the cocoa crop. Despite the fact that they also harvested rice, beans, corn and some tubers for self-consumption, as well as bananas that they sold every two weeks, cocoa was their main source of income. Consequently, many men had to leave the community to work on banana plantations.

> Before, they extracted their cocoa to buy what was necessary. The rice, the beans, the corn, they produced it. But more than anything, with the cocoa we bought oil, salt, candles, matches, clothes, shoes, that is what we bought. But after the *monilinia* that was the difficulty, because people did not have anything for buying. From then on money was scarce, then people emigrated to the banana plantations of La Chiquita in Cahuita, in Sixaola. Most of the young ones, 19, 20 years, went there to work. (BM01, partner of Stribrawpa, personal communication)

This had a huge impact on the life of the community because of the separation and lack of attention to families. Moreover, the labour in plantations increased alcoholism problems and the intensive use of agrochemicals provoked multiple health problems.

> I worked fumigating banana, worked several months and then came back. I came back because the poison was having its effects already. I felt burning sensations on my skin. One fumigates and the wind brings it back to the face, to the body and also to the eyes. Then I had to quit. (JB02, partner of Stribrawpa, personal communication)

In this context, a group of women in the community, worried about not having economic alternatives, had the idea of creating and selling handicrafts in touristic places. The initiative was not well regarded by some men, who did not like women leaving their homes to go and sell things or to participate in training workshops.

> It was the time of a hard fight because the Bribri men were used to their women being good just to stay home, take care of their children, and help them in the mountains, not to be in the organization. So men did not want them to organize, they thought men were the only ones that had the right to organize. (DP01, partner of Stribrawpa, personal communication)

Despite the men's reluctance, the women developed this activity for six years. However, high transportation and living costs in touristic destinations did not make their effort profitable. Therefore, they organized a workshop in the community about alternatives they could launch. That was the moment when they thought of the possibility of developing a touristic offer inside the community, intending to bring the buyers of handicrafts to them, thus avoiding the cost of export. They had previous experience of technicians who had come to the community to work on the cocoa problems and who bought handicrafts while resident. Nevertheless, they had to once again face rejection from some men in the community, who still did not appreciate the potential benefits of this initiative and thought success was not possible.

> The three of us organized a meeting. Many men were leaders, but they did not agree with us because they were not used to seeing us as leaders, but rather seeing us at home, not as leaders, even less seeing a woman in a meeting, in a workshop, no, no. Then they said that they did not agree, that it was in vain and that it would not exist. (BM01, partner of Stribrawpa, personal communication)

The main leader of the group, Bernarda Morales, had been participating in previous years in diverse capacitation processes with different organizations, in terms of leadership, organization or tourism, as well as gender perspectives. While the association incorporated more young people, both women and some men, it kept developing gender perspective training.

In this context, two initiatives competed to receive the support of the Programa de Pequeñas Donaciones (PPD) (Small Grants Programme) of the United Nations Development Programme (UNDP): one on tourism services, boosted by the group of women, and another aimed at reactivating cocoa production, led by a group of men. PPD decided to support the women's touristic proposal and, therefore, the construction of accommodation and a kitchen began. Thanks to a small tour operator in Puerto Viejo they started to receive their first visitors.

The initiative, shaped in 1992 under the name Asociación Indígena Stribrawpa, which means 'Artisan Women' in Bribri, was devised and boosted by three main goals: the rescue of their mother tongue, the improvement of family economy and the protection of the environment. Men can also participate in the project, although the management board is made up of women.

> Men have not been part of the management board because they say we [women] were the fighters that allowed for the organization to exist, and that they were alive and not dying at 50 years old because we were very resilient. They were young and they saw what our fathers said to us and everything, and they say that it was not fair that we had to put up with all those insults. Therefore, they say they do not want to be part of the management board, it is better if women continue. (BM01, partner of Stribrawpa, personal communication)

The association is composed of 40 people, 19 women and 21 men, although not everyone is active at the same time. Once the association was constituted, the assembly determined that new admissions, except the sons and daughters of the ones that were already members, had to pay 500,000 colons (about US$1500) over five years. Until the whole payment is complete, the person does not have full rights in the association.

Stribrawpa was also present in the process of the creation of the association ACTUAR (Cañada, 2017b), in which the main leader was the vice-president and president of the management board. Being part of ACTUAR helped them with promotion and with regularly receiving tourists, in addition to learning how to improve the quality of the services offered and the management procedures. This allowed them to directly work with diverse tour operators, which granted their functioning when ACTUAR dissolved at the end of 2018.

Nowadays, the main income of the community comes from tourism. Between 2014 and 2018, annual visitor arrivals oscillated between 1200 and 1300 tourists. Approximately half of these reside overnight in the community (one or two nights, maximum) and the other half are day visitors. There are six lodgings with a capacity for 30 people, and two dining rooms inside the community. Furthermore, they have prepared a programme to receive volunteers, who stay for a while in the community and bring support for different communal tasks or, if they stay for a long time, may collaborate in the administrative management of the initiative. In these cases, the volunteers live in families' homes instead of staying in the collective accommodation. The vast majority of tourists are foreigners from France, Germany, Spain, the USA and Canada. They arrive at Yorkín mainly through tour operators and, in a few cases, directly through their Facebook profile, but always with a booking.

The main attraction for tourists, in addition to a canoe excursion on the Yorkín river from Puerto Bambú, which takes between one and two hours depending on the level of water, and the season, is participation in diverse cultural activities, such as learning how to make chocolate, how to weave or hearing the stories of the Bribri community.

The work linked to tourism is organized in different roles and rotations. In the different tasks, women and men participate, except for boatmen.

> There are boatmen, cooks, guides, waiters, those who prepare the beds, cleaners, those who develop cultural activities. We all work, all 35 of us, by turns … By turns means that, for example, if we are seven guides, today is the turn of one person and tomorrow another one, and this is how we all work equally. (DP01, partner of Stribrawpa, personal communication)

The payments for each area are not fixed, they depend on the number of tourists attending and on the effort that is considered necessary, with tariffs established in the association's assembly. The workers who receive more rewards are the boatmen, cooks and guides and the ones who receive less are those related to cultural activities.

After tourism, the following sources of income by order of importance are: cocoa, which is now harvested again with varieties more resistant to *monilinia*, and with better maintenance; and bananas, which are cut every two weeks and sold in Puerto Bambú. At the same time, every family produces basic seeds and other tubers and fruits in their own vegetable garden for self-consumption.

To avoid excessive dependence on tourism, the community made the decision that everyone involved in the activities of the association had an obligation to produce food in their own vegetable garden and sell part of this harvest to the association.

> We do have tourism, but not with the idea that it should be only tourism, just tourism and forget about everything else. In our minds this is not it, it is about looking for just a part of something, of a mean, and the other part is agriculture, production. Therefore, we do not wish for 50 people every day because that would have destroyed all of this, it is just a mean. (EMA02, partner of Stribrawpa, personal communication)

They have also prepared control measures through their internal regulations to act against bad behaviour, especially that of men. For instance, being disrespectful with a female co-worker, being violent with their partner, spending income on alcohol or neglecting their family. If someone breaches these regulations they can be reprimanded.

> All these regulations are read to the person when they arrive and they have to sign [a document] expressing that they agree and they are willing to obey them. We have done it because of liquor, basically because of liquor. They have benefits, but not for spending them in sugarcane liquor (*guaro*), but rather for the benefit of their family. (BM01, partner of Stribrawpa, personal communication)

At the same time, and intending to avoid the possibility that tourist success in the community could lead to the arrival of a huge number of people from other communities looking for a job opportunity, they decided to support the launching of other tourism initiatives and agreed that they did not wish to exceed 1500 tourist arrivals per year.

> See, with 1500 people we are fine because we do not want to be millionaires, we want to live better, that is the word, no lack of rice for families, that we are all fine. (CSM02, partner of Stribrawpa, personal communication)

Families that are not in the association receive diverse indirect benefits for the presence of tourism in the community, either from the purchase of food in the community or the creation of communal infrastructures and equipment, as in the case of the high school.

> Now there is work in the community, people do not leave because there are no jobs … Culture has been rescued. Before, for example, nobody wanted to consume the natural things produced here. This has changed a little. You see, the school, our houses, everything is done the way our culture is. (DP01, partner of Stribrawpa, personal communication)

3.5 Discussion and Conclusions

The three cases of community-based tourism analysed here share the fact of having consolidated their commercial strategies. This has allowed them to increase their incomes, to bring economic stability and improve the living standards of their members. It has also generated relevant environments and experiences that produce a notable degree of satisfaction among visitors. Moreover, they have developed each initiative by exploring the best forms of organization of touristic activity that promote strong local participation and a willingness to control the functioning of a part of the touristic activity's chain of value. From this perspective, the three cases show that tourism degrowth for this type of community multiplies the redistributive capacity of tourism and creates more satisfactory and healthy experiences, both for the tourists and the hosts.

At the same time, the three initiatives have developed strategies that are not that common in the majority of community-based tourism initiatives and which make them even more relevant as laboratories of post-capitalist transition processes inside a degrowth logic. It is precisely the analysis of these particular practices that becomes more necessary in the current context, in which we need to find clues to build real utopias to capitalism – once again quoting Erik Olin Wright – in order to implement real changes in the tourism field.

In the first place, the selection of markets is significant. In two of these initiatives, Cooperativa Los Pinos and Ecoposada El Tisey, they have opted to boost their proximity market, composed of middle and lower middle classes. There can also be the presence of national residents who live abroad and come back for vacations or a small quantity of foreigners. The choice of proximity markets, usually poorly valued in the policies of support to the sector, shows diverse potentialities, such as reducing vulnerability in relation to the uncertainty of international tourism in situations of crisis or emergency. At the same time, it allows for the consolidation of a food market in the community apart from what is consumed in the initiative. Additionally, this type of proximity tourism reduces environmental costs, compared to international tourists. Furthermore, it democratizes access to leisure in natural surroundings for the local population. The other analysed initiative, Stribrawpa, has consolidated with a more conventional market of foreign visitors, partially due to the access difficulties which increment transportation costs. However, they have established corrective mechanisms to prevent the risks that could come from higher growth. Therefore, it is significant that volunteer limits have been set to the number of tourists expected, and so they give support to other initiatives to be able to respond to the tourism flows interested in the area and the Bribri culture.

A willingness not to depend exclusively on tourism is notable in all three cases, leading to a diversification process that complements rural economies. Tourism appears after agricultural and livestock activities and initially is seen as a support, but, although it consolidates, it does not imply a *replacement* process in any of the cases. Contrarily, tourism is used to potentiate production and the sale of agricultural and livestock products, generating new markets that are more profitable due to closer proximity with the final consumer, as the cases of Cooperativa Los Pinos with coffee, or Ecoposada el Tisey with vegetables, dairy products, flowers, plants or coffee illustrate. Furthermore, in the case of Stribrawpa, there exists the creation of institutional mechanisms that regulate the obligatory production of food. This option for the link of tourism with agricultural and livestock production helps reduce the potential vulnerability of some of these activities.

It is especially interesting to analyse the adopted organization forms: cooperative, broad family and association. This proves the potential of diverse forms of collective organization to develop tourism initiatives under the control of the local population. One of the key factors was that in the three cases there has been local leadership and they have avoided dependency on external actors. The strengthening of capacities has been of crucial importance in this respect. Through different institutional mechanisms, the three experiences have also been able to manage disagreements and conflicts, and to impose mechanisms of social control, in some cases through the consensus generated in family structures and, in other cases, in a more regulated and institutionalized way.

The community-based tourism initiatives analysed show a diverse repertoire of ways of organizing touristic activity, with more satisfactory results for everyone involved than the ones offered by conventional capitalist structures. Despite this, community-based tourism is not free of risks and contradictions. Currently, there is the possibility that these kinds of initiative might be subordinated to the commercial dynamics of a post-Fordist tourism market, which needs new products and experiences (Cañada and Murray, 2019). This can lead to a reduction of the control capacities of the local population in comparison to big corporations. Therefore, it is necessary to identify these threats vs the risk of idealizing all types of community-based tourism. However, some experiences allow thinking of community-based tourism as a contributor to transforming tourism from a degrowth perspective. This opens the door to widening an investigation agenda focusing on the comparative analysis of these kinds of initiative as post-capitalist trials of production organization and tourism consumption.

The global crisis due to COVID-19, officially declared a pandemic by the director of the WHO on 11 March 2020, has had among other effects the almost total stoppage of tourism around the globe. Until there is a vaccine with sufficient coverage, international tourism will be severely restricted or subjected to strong health control measures. The possibility of some situation of crisis reducing international tourism activity has become increasingly feasible. The causes are multiple: conflicts or security issues, higher incidence of natural phenomena or due to health alerts such as the current one. This means that tourism initiatives in many territories are more vulnerable. In this context, it is probable that tourism activities that may be developed in the short and middle term are fundamentally linked to proximity. Having a strong base in the national and proximity market allows for territories and tourism initiatives to be more resilient. In the case of Nicaragua, as explained above, this behaviour is seen in community-based tourism. Therefore, the communal tourism initiatives that could better face the sociopolitical crisis started in April 2018 were the ones that had consolidated a proximity clientele, such as the Ecoposada El Tisey. The situation is different for experiences dependent on the international market, since they almost had to close their businesses. In this context, one of the pillars of degrowth, which aims to reduce travel and power consumption, shows more capacity of adaptation to tourism centred on proximity. Despite this, degrowth effects are not the same when it is due to a forced phenomenon, which leads to social disaster, or when it is done voluntarily. The possibility of a successful community-based tourism as a business experience, based on proximity tourism or international tourism with limitations, as seen in the three case studies in this text, shows possible paths to degrowth that make their communities less vulnerable to global crisis such as the current pandemic.

Notes

1 Ernest Cañada is the coordinator of Alba Sud, an independent centre of investigation specialized in responsible tourism and professor of the Barcelona School of Tourism, Hospitality and Gastronomy of Barcelona University (CETT-UB).

2 This text has been elaborated in the framework of the project 'Research Platform on Tourism, Human Rights and Gender Equity' developed by Alba Sud with the support of the Catalan Agency for Development Cooperation (ACCD) (call 2019).

References

Andriotis, K. (2014) Tourism development and the degrowth paradigm. *The Business of Tourism* 13, 37–46.

Andriotis, K. (2018) *Degrowth in tourism. Conceptual, theoretical and philosophical issues*. CABI, Wallingford.

Blanco-Romero, A. (2019) Tourism degrowth. In: Cañada, E. and Cañada, E. (eds) *Tourism in the geopolitics of the Mediterranean*, Contrast Reports Serie, no. 9. Alba Sud, Barcelona, pp. 66–69.

Blanco, A., Blázquez, M. and Noves, G.C. (2019) Resiliencia territorial y turística en un mundo polarizado. In: *Crisis y espacios de oportunidad. Retos para la Geografía. XXVI Congreso de la Asociación Española de Geografía. Libro de Actas. Valencia, del 22 al 25 de octubre de.* 2019. AGE and Universidad de Valencia, Valencia, pp. 663–677.

Blázquez-Salom, M., Blanco-Romero, A., Vera-Rebollo, F. and Ivars-Baidal, J. (2019) Territorial tourism planning in Spain: from boosterism to tourism degrowth? *Journal of Sustainable Tourism* 27(12), 1764–1785.

Burrai, E., Buda, D.-M. and Stanford, D. (2019) Rethinking the ideology of responsible tourism. *Journal of Sustainable Tourism* 27(7), 992–1007.

Cañada, E. (2014) *Turismo comunitario en Centroamérica. Experiencias y aprendizajes*. Editorial Enlace, Managua.

Cañada, E. (2017a) Contribución del turismo comunitario a la economía campesina: la Cooperativa Los Pinos en El Salvador. In: Gascón, EnJ. and Milano, C. (eds) *El turismo en el mundo rural. ¿Ruina o consolidación de las sociedades campesinas indígenas?* PASOS, Colección PASOS Edita, 18, La Laguna (Tenerife), pp. 111–137.

Cañada, E. (2017b) Estructuras de intermediación turística procomunitarias. La experiencia comercial de ACTUAR en Costa Rica. *Gazeta de Antropología* 33(1). Available at: edición online: http://www.gazeta-antropologia.es/?p=4973

Cañada, E. (2018) Dispossession, displacement and subordination in the construction of tourist areas: central America as a conflict scenario. *Norois* 247, 49–62.

Cañada, E. (2019a) Los mercados del turismo comunitario en América Latina. Perspectivas para una agenda de investigación. *Dimensiones Turísticas* 5(3), 96–105.

Cañada, E. (2019b) *Transformaciones en las relaciones de género en experiencias de turismo comunitario en Centroamérica*. 7. Barcelona, Alba Sud Editorial, Informes en Contraste.

Cañada, E. and Murray, I. (2019) Introducción. Perspectivas críticas en turismo. In: Cañada, E. and Murray, I. (eds) *Turistificación global. Perspectivas críticas en turismo*. Barcelona, Icaria Editorial, pp. 7–34.

Chassagne, N. and Everingham, P. (2019) Buen Vivir: Degrowing extractivism and growing wellbeing through tourism. *Journal of Sustainable Tourism* 27(12), 1909–1925.

Clancy, M. (2017) *Slow tourism, food and cities. Pace and the search for the "good life"*. Routledge, London.

de Bloom, J., Nawijn, J., Geurts, S., Kinnunen, U. and Korpela, K. (2017) Holiday travel, staycations, and subjective well-being. *Journal of Sustainable Tourism* 25(4), 573–588.

Demaria, F.K., Kallis, G. and Bakker, K. (2019) Geographies of degrowth: Nowtopias, resurgences and the decolonization of imaginaries and places. *Nature and Space* 2(3), 431–450.

Egresi, I. (2016) Post-Fordism, alternative tourism and sustainability. In: Egresi, I. (ed.) *Alternative Tourism in Turkey*. Springer, New York, pp. 271–283.

Fletcher, R., Murray, I.M., Blanco-Romero, A. and Blázquez-Salom, M. (2019) Tourism and degrowth: an emerging agenda for research and praxis. *Journal of Sustainable Tourism* 27(12), 1745–1763.

Flores, M., Bratescu, A., Martínez, J.O., Oviedo, J.A. and Acosta, A. (2002) *Centroamérica: el impacto de la caída de los precios del café*. 9. México DF, CEPAL, Serie Estudios y Perspectivas.

Gascón, J. (2018) Turismo, gastronomía y alimentación: de la teoría del enlace a la patrimonialización de la gastronomía. In: Medina, F.X. and de Leal, P.M. (eds) *Gastronomía y turismo en Iberoamérica*. Ediciones Trea, Gijón, pp. 15–32.

Gómez, R. and Pardo, D. (2019) Southern Europe as an anti-tourism political space. In: Cañada, E. and Cañada, E. (eds) *Tourism in the geopolitics of the Mediterranean*, Contrast Reports Series, no. 9. Alba Sud, Barcelona, pp. 86–91.

González, F. (2018) Coastal tourism in a context of neoliberal development: social change in Mancora, Peru. *Bulletin of Latin American Research* 39(2), 1–15.

Gössling, S. and Peeters, P. (2015) Assessing tourism's global environmental impact 1900–2050. *Journal of Sustainable Tourism* 23(5), 639–659.

Heimtun, B. (2017) Home holidays as real holidays? Midlife single women's experiences. In: Khoo-Lattimore, C. and Wilson, E. (eds) *Woman and Travel. Historical and Contemporary Perspectives*. Apple Academic Press, New York, pp. 202–215.

Higgins-Desbiolles, F., Carnicelli, S., Krolikowski, C., Wijesinghe, G. and Boluk, K. (2019) Degrowing tourism: rethinking tourism. *Journal of Sustainable Tourism* 27(12), 1926–1944.

Ioannides, D. and Debbage, K. (1997) Post-Fordism and flexibility: the travel industry polyglot. *Tourism Management* 18(4), 229–241.
Ioannides, D. and Debbage, K.G. (1998) Neo-Fordism and flexible specialization in the travel industry. In: Ioannides, D. and Debbage, K. (eds) *The Economic Geography of the Tourist Industry: A Supply-side Analysis*. Routledge, Abingdon, pp. 97–119.
Izcara, C. and Cañada, E. (2020) Staycation: ¿una forma de entender el turismo de proximidad? *Alba Sud* 17/04/2020.
Kousis, M. (2000) Tourism and the environment: a social movements perspective. *Annals of Tourism Research* 27(2), 468–489.
Latouche, S. (2006) *Le Pari de la décroissance*. Fayard, Paris.
Latouche, S. (2007) *Petit traité de la décroissance sereine*. Mille et Une Nuits, Paris.
Marín, G. (ed.) (2015) *Sin tierras no hay paraíso. Turismo, organizaciones agrarias y apropiación territorial*. El Sauzal, PASOS.
Milano, C. (2018) Overtourism, malestar social y turismofobia. *Un debate controvertido. PASOS. Revista de Turismo y Patrimonio Cultural* 18(3), 551–564.
Milano, C., Novelli, M. and Cheer, J.M. (2019) Overtourism and degrowth: a social movements perspective. *Journal of Sustainable Tourism* 27(12), 1857–1875.
Moltz, J. (2009) Representing peace in tourism mobilities: staycations. slow travel and the amazing race. *Journal of Tourism and Cultural Change* 7(4), 270–286.
Oh, H., Assaf, A.G. and Baloglu, S. (2016) Motivations and goals of slow tourism. *Journal of Travel Research* 55(2), 205–219.
Ojeda, D. (2016) Landscapes of dispossession: proposals for an analysis from the entailed sociospatial reconfigurations. *Revista Colombiana de Antropología* 52(2), 19–43.
Özdemir, G. and Çelebi, D. (2018) Exploring dimensions of slow tourism motivation. *Anatolia* 29(4), 540–552.
Renkert, S.R. (2019) Community-owned tourism and degrowth: a case study in the Kichwa Añangu community. *Journal of Sustainable Tourism* 27(12), 1893–1908.
Ruiz-Ballesteros, E. (2017) Socio-ecological balance in community-based tourism experiences: a research proposal. In: Butler, R. (ed.) *Tourism and Resilience*. CABI, Wallingford, pp. 41–452.
Scheyvens, R. and Biddulph, R. (2018) Inclusive tourism development. *Tourism Geographies* 20(4), 589–609.
Serdane, Z., Maccarrone-Eaglen, A. and Sharifi, S. (2020) Conceptualising slow tourism: a perspective from Latvia. *Tourism Recreation Research* preprint.
Timms, B.F. and Conway, D. (2012) Slow tourism at the Caribbean's geographical margins. *Tourism Geographies* 14(3), 396–418.
Wright, E.O. (2010) *Envisioning real utopias*. Verso, London.

4 Degrowing the Commoditization Process in Community-based Tourism and Local Entrepreneurship

Ammalia Podlaszewska*

Bauhaus-Universität Weimar, Weimar, Germany

4.1 Introduction: Alternative Growth

The hypothesis of degrowth posits that such a trajectory of social transformation is necessary, desirable and possible. From the point of view of so-called economic stability, degrowth signifies 'radical political and economic reorganization leading to drastically reduced resource and energy throughput' (Kallis *et al.*, 2018, pp. 291–316), mobilized within the jargon *decroissance* (degrowth) initially in the late 1960s and early 1970s. The depth of internationalization becomes more apparent in the environmental justice movement that challenged the growth paradigm as a critique of over-consumption and over-production. It was further activated by environmental campaigners in Lyon, France, at the beginning of the 2000s (Demaria *et al.*, 2013). The argument of degrowth has questioned the limits to growth debate with critical examination of the transformations that have made economic growth a dominant narrative.

During the Leipzig International Conference in 2014, scholars started to cross-check the concepts and tools across disciplinary boundaries and beyond, to investigate 'condition[s] for the emergence and proliferation of visible alternatives' (Degrowth Leipzig Conference, 2014). In the article, 'The Economics of Degrowth'', it is similarly argued that the conference drew attention to values and political systems that promote involvement of communities in the struggle for the production of knowledge (Kallis *et al.*, 2012, pp. 172–180). The conference reached a common understanding that supports

*ammalia.podlaszewska@gmail.com

DOI: 10.1079/9781789245073.0004

conviviality and social changes towards efficiency. The editorial team of the conference proposed:

> Degrowth does not only challenge the centrality of GDP as an overarching policy objective but proposes a framework for transformation to a lower and sustainable level of production and consumption, a shrinking of the economic system to leave more space for human cooperation and ecosystems.

Research on degrowth is becoming more diversified within themes and disciplines that come from political science, economics, ecological economics, history, anthropology, social science and technology. The variety of background proves that degrowth as a normative concept has analytical and practical applications. Moreover, the correspondence debates on degrowth between the Global North and post-development in the Global South have acknowledged that inequality in degrowth impedes the cultural and historical differences between northern and southern countries and their societies with different development phases. Many alternative movements to growth are seeking to redefine the relationship between the two while challenging productivism and resisting the homogenization effect of economic and cultural symbolism (see Latouche, 2009; Martínez-Alier *et al.*, 2010; Martínez-Alier, 2012; Demaria *et al.*, 2013). Andriotis (2000) further argued that the consideration of the needs of the local community is often missed in addressing the importance of environmental and cultural conservation in development. This movement calls for more diversity of context between plural movements rather than imposing a homogeneous model for transformation. In the literature study by Weiss and Cattaneo (2017), the academic discourse of degrowth is dominated by authors from European countries with a clear majority of contributions from Spain, with the exception of studies from Colombia and Cuba (see Boillat *et al.*, 2012 and Escobar, 2015).

Research into degrowth in the Global South raises numerous questions such as 'what kind of diversified policies and on what scale?' are appropriate for the Global South's development context. The tension between degrowth and transition discourses has reinforced the significance movement which brings the dominant societal structure into question. Degrowth is seen as 'a concern of elites' who are more middle class and who do not know poverty from their own experience (Rodríguez-Labajos *et al.*, 2019, pp. 175–184). The understanding of this idea requires a context of overall welfare and over-consumption. For many people in the Global South, degrowth is not appealing because society '[has] suffered from situations of poverty and scarcity of the most basic needs', therefore 'some growth to reach more security in terms of survival is regarded logical' (Rodríguez-Labajos *et al.*, 2019). Although degrowth is not the equivalent of 'negative growth' or 'zero growth', research on degrowth from a social science perspective also draws insight from ethnographic social systems that have managed well without growth. The growth hegemony can also be measured by searching for alternative indicators to resist the Keynesian-led economy that was criticized for its interdependence on consumerism and its commoditization imaginary (Latouche, 2009).

In this chapter, the articulation of degrowth will be narrated through the practices of community connection, comprising grass-root initiatives and networks. The chapter asks what are the equivalences between core themes in the degrowth debate in the Global South that have exacerbated the existing forms of social and spatial equality in providing alternative capital opportunities for communities. The chapter aims to discuss relevant constructs of commoditization in community-based tourism to point out the key features of alternatives in tourism planning within the degrowth paradigm.

The case study focuses on the city of Bandung, the capital of West Java, Indonesia. As a developing country, Indonesia still perceives tourism as a means of enhancing its economic growth and its political legitimacy (Richter, 1992, pp. 35–46), after all, as one of the growing pillars of Indonesia's economic growth, the tourism industry in 2009 was ranked as the third most important sector after gas and oil. Bandung's tourism is one of the main stimuli of new entertainment facilities and urban infrastructure in the city. Tourism in Bandung is modernized through cultural heritage tourism, moving towards a more consumptive pattern such as shopping activities, and gives a significant contribution to the regional economy. The dynamic modernization that is growing today in Bandung includes the creative industries. Its development has become a factor that strengthens the tourism infrastructure of trade, hotels and restaurants, as well as the services and manufacturing sector.

4.2 Tourism in a Degrowth Paradigm

The diversification of consumption activities in the urban landscape is referred to as 'consumption of leisure'. Cities provide the context for a diverse range of social, cultural and economic activities where tourism, leisure and entertainment form a postmodern society. In tourism, the relationship that occurs between space and human experience is valued by the occurrence of mutual dynamics: the city adapts tourism as a reference model of development, and tourist products adopt urban dynamics to meet the demands. In the growth perspective, tourism tends to appropriate the spatial outcome of global capitalism on the local level and the spatial organization brought by urbanization. The consequences are reflected in the humanization of the service industry after the critique of mass tourism (Fordism to post-Fordist era) which was also influencing the nature of urban planning in general. At the same time, city productivity takes a key role as the main feature of economic growth of a city; it profiliates the planning for improving labour and capital efficiency and integrating land use and infrastructure (Podlaszewska, 2017).

Tourism and growth are synonymous with the notion of transformation which is used as a synthesized concept to express the importance of space and time. Considering these relevant concepts, tourism is treated as a catalyst for the transformation of the economy, morale and the appearance of a city. In areas where local commoditization seems possible, tourism is used as a vehicle which only came into view at a later stage of the urban planning process. While there is definitely a set of preconditions which need to be assessed in order to give this catalyst function, it is therefore necessary to understand the policy context and spatial functional context and their implications to the community. Some bodies of literature link urban tourism to the postmodern city (see Urry, 1995; Page and Hall, 2003; Selby, 2004). The postmodern movement has arrived to emphasize the cultural dynamics of society, utilizing the service industry as part of a fascination with consumerism.

The degrowth paradigm in tourism research has come to the fore in respect of finding alternatives to the perceived environmental, social and economic down-spiral of traditional mass tourism (Kallis *et al.*, 2018). Tourism studies from globalization studies (Sassen, 1991; Fainstein and Judd, 1999) point out its importance, mostly derived from the perspective of a 'central place' paradigm, with reference to the concept of the global city, located in global circuits of money and people with the status of cultural centre and global business network. The alternative to tourism, on the other hand, emphasizes as its main argument the 'localized' and community-based perspective. It focuses on how capacity can be built for and from tourism to enhance the influence of community on urban planning.

The growth of the tourism industry and its economic potential for stimulating local entrepreneurship has formed an important point in much of the literature on tourism (De Kadt, 1979; Shaw and William, 1990; Shaw and William, 1998).Within the considerable literature mentioned above, in regard to the economic potential of tourism, little attention has been paid to the role of entrepreneurial activity and how tourism enterprises operate in different economies (Shaw and William, 1990, 1998).

Degrowth in the context of tourism could highlight the benefit of collaborative community development and the attempt to de-commoditize local attributes merely as a symbolic economy. It should be noted that the motive of tourism degrowth is not anti-tourism, rather a call to descale the industry's negative impact and allow the optimization of benefits to the local community and ecosystem.

4.3 Spatial Embeddedness of Degrowth

While developing rapidly, Southeast Asian countries are still facing a challenge within their institutional and societal structures. The pattern of urbanization in large Southeast Asian cities is usually analysed in terms of natural increases in the population who already reside in the city, migration of people to urban from rural areas and the expansion of the city (administratively and spatially) from non-urban areas into urban areas (Jones, 2002; Podlaszewska, 2017).

More than 11% of Southeast Asia's population has been residing in the extended metropolitan regions (McGee *et al.*, 2007), representing a form of urbanization that has not been sufficiently developed, yet has been profoundly affected by urbanization. The term *desa kota*, referring to regions with an intense mix of agricultural and non-agricultural activities that often stretch along a corridor between large city cores, literally means village (desa) and city (kota) in the Indonesian language (Hugo, 1996). On the other hand, the term *kampung kota* (urban *kampong*) is often associated with a dense population settlement in the urban core area whose population is often regarded as the working-class serving local economic activity (households, small traditional markets, sellers, drivers, servants, waitress, etc.). As a major component of economic activity, although the inhabitants of urban kampongs are not working in the agriculture sector any longer, the community connection is reflected with a kinship relation. From this perspective, urbanization in Indonesia has formed a new spatial outcome and community reorganization.

To contextualize the equivalences of core themes in the degrowth debate in the Global South, this chapter concentrates on places that are embedded in the built environment of the city, especially in city neighbourhoods (or urban kampongs) and communities. Urban kampongs are referred to as attributes of locality and identity, where the ongoing interaction of everyday life patterns between people, activities and territory, and community resources and identity, is a daily practice (Murti, 2015). These kampongs are typically located within the proximity of the city centre with access to a labour supply and an economic centre, however they tend to have limited accessibility to urban services and technical provision for basic infrastructure. It is relevant to point out that this inherent condition has strengthened the organization of the community, where its connection defines cooperative bonds that are culturally imbued. On the other hand, urbanization seems to weaken the primary bonds of urbanites to neighbor and kinship, where its structure has challenged the community connection of urban kampongs. The ethnographic studies of neighbourhoods by Willmott and Young (1960) have provided a basic view on community life, understood in terms of processes other than those involved in the broad ecological structuring of modern settlement patterns.

In the Indonesian context, 'community' life is vividly found in urban kampongs – that is, the fundamental connection between the city and its people (Sihombing, 2004). From an administrative perspective, community in urban kampongs is a concept of social relationship (*rukun*), embodied in a community institution (*rukun tetangga*) as an organizational unit associated with the kampong as a highly unstructured and informal settlement (Firman, 2002). The difference between urban *kampongs* (kampung kota) and village kampongs (kampung desa) is relevant to point out since the latter are influenced by trade and commercial goods and thus have a high population density. This proves that the presence of urban kampongs is highly influenced by the presence of the city as the centre of political and economic activity (Geertz, 1965).

The relationship of an urban kampong with the city is also shaped by its link to the broader urban spatial economy and political power that are utilized to secure access to land and housing or infrastructure (Sihombing, 2004). This link embodies the elaborated knowledge of the informal sector that is fundamentally intertwined with the informal practices of the community. Since the community serves as the operator of the informality, its local resources represent the commodity value. The term kampong will be discussed in this chapter as the equivalent to urban slums and nostalgia. As the utmost overrated principle of kampong tourism, the romanticism over poverty and poor living conditions is associated with slumming, while the valorization of marginalized urban areas seems to be under-investigated (Freire-Medeiros, 2009; Mowforth and Munt, 2009; Butler, 2012). While it is evident that most tourist attractions are immersed within the historical quarter of the city, by the same token kampong tourism also manifests itself in the middle of areas of relative deprivation.

4.4 Commoditization and Tourism

Urry (1995) considers how places have been transformed by the development of service occupations and tourism industries. For many cities, as is argued by Castells (1996), urban growth is accentuated by and mostly depends upon services, especially the more advanced services, which deal with the exchange and use of information as the major form of production. This is because being a form of service industry, tourism retains the idea of trade with the objective of revenues. The products provided for tourists by a city shape what we call *urban experience* (Harvey, 1989).

Commoditization processes thus illustrate the valorization of places and local attributes that are restructured as a means of consumption, especially where people can actually experience the social relation (both visitors and locals) to obtain goods and services (Podlaszewska, 2017). It is relevant to navigate the commoditization process in the relation between postmodernism and consumption in the city that might be appearing, as if postmodernism puts the consumer into the main chain of production. Similarly, tourism amenities are offering great opportunities for consumption behaviours for the city dweller. Contrary to the traditional form of production, tourism extends production. Its raw material embodies a hybrid combination of socio-cultural, natural and physical features in which its attributes spread over the locality (Husbands, 1981, p. 50). Commoditization of locality in tourism enables a dependency upon the exploitation of communities and local cultures for their image as a processes of tourism production (McCannell, 1992) because locality is specifically appropriated to maintain the never ending cycle of production and consumption as functional spatial commodities (Bartling, 2007).

4.4.1 Community-based Tourism and Entrepreneurship

The concepts of community-based tourism and small-scale enterprise are synonymous while also perceived as processes aimed at creating social and economic progress for the community through active participation and the initiatives of community members themselves. Community members are seen not as a problematic burden but as a unique potential that has not been fully developed (Rothman, 1968). The importance of building partnership between community and local government and between community and tourism business highlights the nature of community-based tourism (Jamal and Getz, 1995). This model of development became popular in the mid-1990s, reversing bottom-up development approaches in an effort to provide real and all-inclusive community participation at all levels of development (Asker *et al.*, 2010).

While it is considerably relevant that capitalists search for new sources of cultural symbolism to maximize their profit into commodities, the role of local entrepreneurship on the other hand is seen as more socially responsible (Shaw and William, 1998). Much employment in tourism is based on self-employment since tourism is basically a service industry requiring the notion of small-scale entrepreneurship. The variety of local richness, which is embedded in the interaction between cultures and economies, is important to the understanding of capitalism and small-scale industrialization. Small-scale entrepreneurship, according to Jafari (1989) operates according to a tourism business culture, and yet it may also be part of the local social system (Shaw and Williams, 1994, p. 132).

Small-scale enterprises have become a visible and important element in community-based tourism (Dahles, 1997). The 'petty modes' of production continue to increase and diversify as part of the informal sector (Dahles, 2001, p. 94). The basic feature of kampong tourism is that it is defined as the opposite of government-supported 'quality' tourism and is considered as small scale, low budget, unorganized, uncontrolled, informal and of low quality (Guinness, 1989). It also entails more intensive tourist–local interaction, generates local benefit, makes use of local facilities and involves family businesses and self-employed people.

The growth of the tourism industry and its economic potential for stimulating local entrepreneurship has formed an important point in much of the literature on tourism (see Young, 1973; De Kadt, 1979; Mathieson and Wall, 1982; Mcintosh and Goeldner, 1986; Shaw and William, 1990). The rediscovery of the entrepreneurial role in society itself has gained more constant interest across social science (Aldrich, 1999; Swedberg, 2000; Kim and Aldrich, 2005; Ghezzi, 2012). Its increasing popularity has benefited from the hegemonic expansion of neoliberal ideology worldwide (Harvey, 2005).

The categories of small entrepreneurs distinguish themselves by strategically operating networks instead of land, equipment or funds. The tourism-related sector is particularly explanatory of the multifaceted economic relationship that incorporates formal and informal modes of employment. Jobs generated by the tourism industry involve formal and informal sector activities, as the tourism sector continues to grow and diversify (Cukier, 1996). Dahles further emphasized that the combination of formal (mode of employment, seasonality, market demand, etc.) and informal tourism (street guides, vendors, merchandise business, etc.) sectors stimulates small-scale local entrepreneurship and enhances service between formal and informal activities. Thus, the boundaries between the two activities in the tourism industry are blurred at both the individual and at the collective level (Dahles, 2001).

The institutionalization of community in formal formations has been demonstrated through the establishment of local governance institutions at the village level, the level of neighbourhoods, and the small-scale neighbourhood level. The formal institutions of governance are accompanied by supporting institutions in the control of development implementation, such as the BKM (Community Self-help Bureau), Korperasi (Community Economy Union), *karang taruna* (youth clubs) and the PKK (Family Welfare Movement). It can be said that these administrative agencies hold a mediation role. These institutions are constitutionally able to form a bridge between government policy and the aspirations and needs of the community. Non-formal community-based organizations can be categorized as social organizations of non-kinship and kinship. While the latter would not normally appear in communities that have urban characteristics, urban kampongs prove otherwise.

This is related to the level of complexity and specialization of labour and a high level of heterogeneity. While each member is following the rules derived from the cultural values of local communities and establishing a mutual relationship with other members of the organization, it is not obligatory for them to be always involved within particular activities that are not fulfilling their aspirations. Rather, mutual cooperation is the core philosophy of these non-formal organizations. These groups are really rooted in the community; they possess enormous power in mobilizing public support and creating leadership.

The community organization of urban kampongs is reflected in the mutual relationships among members, as well as with the official administrative staff, which is the strength of community connection that can be exploited further. It is also natural that work productivity is dependent upon the leader of the organization; a leader is also the middle-man between government agencies and the private sector which could provide training, education or capital to improve the organization's ability to solve problems and gain know-how.

4.5 Research Methods

The research is based on one case study, which is a method focused on particular characteristics of specific areas (Denzin and Lincoln, 1994). What is typical in an exploratory case study, for instance, is the importance of initial fieldwork and small-scale data collection before proposing research questions and hypotheses, which is crucial in determining the protocol that will be used. In this research, the basis for the theoretical framework of degrowth and tourism is first established to understand the context of commoditization which is presented in the case study of community-based tourism (Fig. 4.1). Following that, the review of the primary and secondary literatures of the social context enabled the formulation of the state of the art of this research.

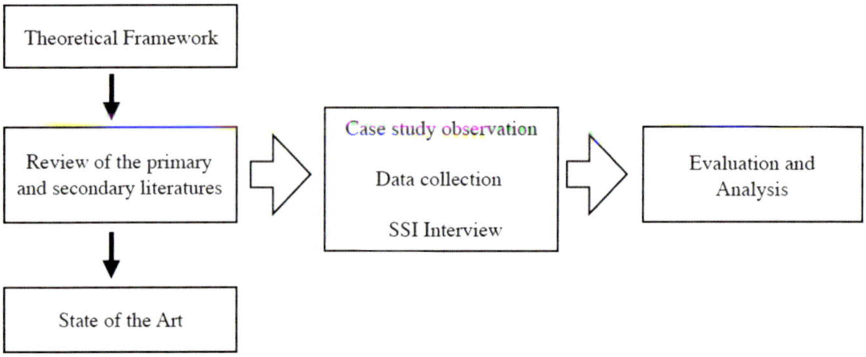

Fig. 4.1. Research framework: tourism and degrowth (author, 2019).

Since the context of this research is constituted by a specific case study, the author is aware of the limitations in analytical rigour and cumulative generalization of knowledge. However, through case studies, it is easier to define best practice in using narrative as an effective tool to handle complex urban processes with unclear boundaries, inputs and outputs.

The aim of an exploratory case study is to explore any phenomenon in the data which serves as a point of interest to the researcher. Meanwhile, descriptive case studies attempt to describe natural phenomena which occur within the data to describe the data *as they appear*. Neuman (2000) suggested that descriptive case studies might be in a narrative form. However, the challenge of a descriptive case study is that the researcher must begin with a descriptive theory to support the description of the phenomenon or else there will be the possibility of lacking reliability. Explanatory case studies examine the data closely both at the surface and at a deep level in order to explain the phenomena in the data.

Case studies examine comprehensively many features of a few cases over time. The researcher uses the logic of analytic induction, considering the context of a case and examining how its parts are configured (Neuman, 2000). Case study method enables a researcher to closely examine the data within a specific context through selecting a small geographical area or a very limited number of individuals as the subjects of the study. Case studies, in their true essence, explore and investigate contemporary real-life phenomena through detailed contextual analysis of a limited number of events or conditions, and their relationships.

Data collection in this research was focused on the village of Rajut Binong Jati, conducted empirically from 27 December 2018 to 9 January 2019, and was carried out through several stages. The initial stage was to observe and recognize the condition of the field and the daily activities of its inhabitants, mostly related to the economic activities in the neighbourhood, such as daily marketplace, street vendor, motor taxi, trash collector, and so on. The second stage was interviewing eight key actors consisting of village leaders from Kampung Rajut, Bandung creative communities and representatives from tourism and trade agencies in order to get first-hand information of related data. The third stage was the collection of data or archives from the internet such as newspaper articles, photos and videos in Bahasa, Indonesia, related to tourism activities in Kampong Rajut. This was carried out parallel to writing this book chapter. After finishing the field research and empirical study in Bandung, the data analysis was conducted, together with online contacts established with the key actors.

Semi-structured interviews were conducted with a relatively small number of key stakeholders. This provides a means to determine how individuals or social groups view themselves, their relationship to the environment, to others and to agents and forces of change. The aim of the selective informant interviews was to gather insightful information about the history of the area, the perceived impact of tourism development, the transformation of the physical environment and community relations, and interviewees' views on participation in the development process. In selecting key informants, the researcher referred to Gotham (2005, 2007) and his study of tourism in New Orleans. The discourse of 'tourism from above' and 'tourism from below', as it is critically conceptualized by Gotham (2005), illustrates the contrasting yet mutually dependent body of tourism organizations responsible for enhancing local distinctive features through a series of strategic branding. The relationship is not contradictory but rather reciprocal and responsive. Tourism from above operates in the process of globalization, commoditization and urban branding, which emphasize the role of regulation strategies established by governance. On the other hand, tourism from below reflects the role of grassroots initiatives to express their cultural invention and local authenticity in everyday life.

Data analysis in this study uses two methods, namely content and descriptive analyses. The content analysis method is useful for comparing theoretical studies related to creative villages and resilient cities and the condition of the data from the research object. The qualitative (descriptive) method is used because the research intends to reveal an understanding of a social problem written in detail. The descriptive method is used to help the writer describe factually the relationships between the instruments studied.

4.6 Descaling Urban Tourism

Departing from the current tourism conditions, Bandung is one of the leading tourist destinations in West Java Province. The development of its tourism is supported by the availability and variety of urban tourism products that are supported by adequate tourism infrastructure and facilities that are interconnected with the webs of street networks and densely populated residential areas (Fig. 4.2). Bandung as a tourist city is also supported by the region (surrounding districts) which is characterized by eight thematic tourism clusters: historical buildings and architecture; art and culture; shopping mall; culinary; amusement park; craft centre and souvenir; natural and landscape tourism; and sport amenities (Department Pariwisata dan Budaya Bandung, 2017).

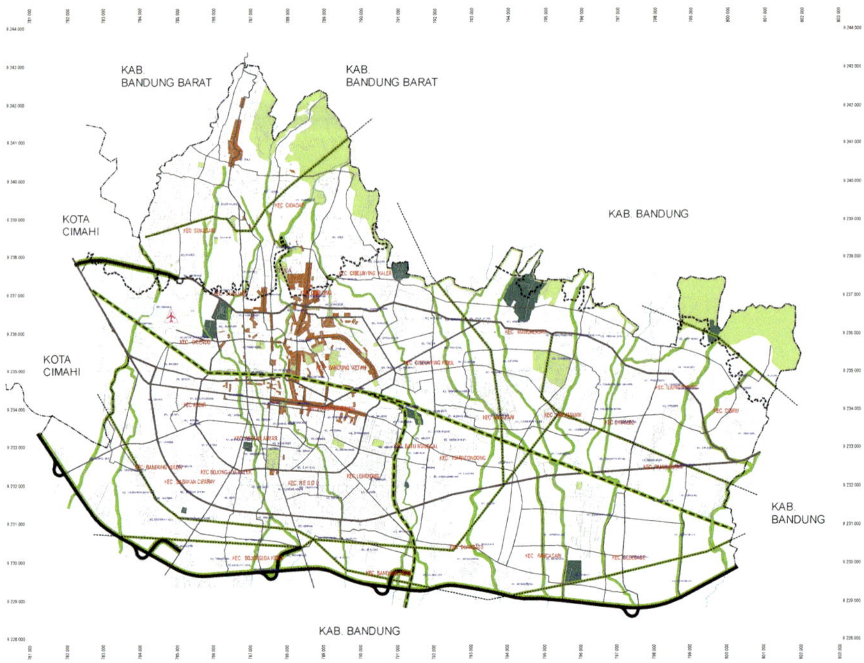

Fig. 4.2. City map of Bandung (Bandung Spatial Planning Report, 2018).

Urban areas with the influence of modern culture and history of Bandung are the main characteristics of this region. Cultural, educational and natural attractions are the leading themes of tourism products in the urban tourism area. Additionally, Bandung is currently dominated by shopping activities, especially with the proliferation of factory outlets, in line with the function of Bandung as the capital of West Java Province and the city of services, MICE tourism products (meeting, incentive, conference and exhibition), as well as education-based tourism (knowledge-based tourism). In addition, Bandung also has capacity as a distribution centre for other tourist destinations in West Java, specifically cities or districts around the city of Bandung. By utilizing its position and potential, many tourism and supporting facilities have grown as attractive tourist destinations.

In the National Seminar on ‘Creative Industries for the Welfare of the Nation’ held in Bandung on 2 April 2008, the Mayor of Bandung at that time, Dada Rosada, revealed the fact that between 2002 and 2006, the number of apparel industries and *distros* (clothing apparel) in the city of Bandung increased from 200 units to 400 units. Along with this development, Bandung’s image as a fashion centre became very strong. Creative industries in the city are divided into seven leading types: 1) apparel/distribution; 2) shoes, sandals, bags and others; 3) electronics; 4) knitting; 5) telematics; 6) automotive from rubber, plastic and automotive spare parts; and 7) creative industries. As a comparison, the development of small- and medium-sized industries in Bandung in 2007 was 10,559 business units, 70,816 workers and

4,541,280,000,000 Indonesian rupiah investment (equivalent to 306,974,179,39 euro) – whereas for large industries, it consisted of 115 business units and 46,686 workers. Overall, the total number of small- and medium-sized industrial units in West Java in 2007 for business units was 10,674, employing 117,502 workers (Badan Pusat Statistik kota Bandung, 2017).

Micro-, small- and medium-sized enterprises (MSMEs) contribute greatly to the Indonesian economy and to ASEAN. Around 96% of business forms in ASEAN are composed of MSMEs with contributions to GDP of around 30 to 57% and absorbing employment around 50 to 95%. While in Indonesia, MSMEs contribute to 99.98% of business units with a contribution of 57% to national GDP and absorb more than 97% of the domestic workforce. The role of MSMEs is clearly very important as the main driver of job creation and economic growth, both at national and regional levels.

Based on data from the West Java Department of Industry and Trade in 2015, Bandung City has the highest number of business units and manpower compared to other cities in West Java Province. The economic growth of the City of Bandung is supported by the textile, processing, trade and services industries (Department Perdagangan dan Perindustrian Bandung Report, 2017). The tourism sector for the city is increasingly becoming a mainstay. In 2014, the city was able to surpass Bali Province for the level of domestic tourist visits. One of the attractions of the city is the concept of the tourism market, which exists in order to accommodate and promote the work of the SMEs in marketing their products (Badan Pusat Statistik kota Bandung, 2017). The city government of Bandung has identified the concept of a destination zone, in which the local textile industry has been packaged and branded as one of the fundamentals of community-based tourism.

Community-based tourism in Bandung is a movement that is supported by the local government, usually through cooperation with other cultural/economic stakeholders through various incentives and capacity-building programmes promoted by the municipality. In a community development programme such as Kampung Kreatif, community participation is fundamental to realize the desired changes in the community. The process of community development in Kampung Kreatif should not be imposed from the outside and cannot be determined by community workers, representative councils or government agencies, because it is a process that is carried out, initiated and monitored by the community itself.

Kampung Kreatif is an important part of creating a resilient and adaptive city. Creative areas (of kampong kota) can easily adapt to environmental changes and survive all kinds of problems that befall them. In addition, people who live in kampong kota are considered as a potential social resource that can be fostered to strengthen the image of the city. Local commodities in Kampung Kreatif are the community's ideas and social, cultural, and environmental conditions of the kampong. The role of the creative village in the city economy is the independence of the population in carrying out creative activities in the arts and tourism.

As an integral part of tourism activities, Kampung Kreatif in Bandung accelerates income distribution, increases employment opportunities, tax

Fig. 4.3. A small shop of knitted products (the author, 2019).

revenue, national income and the added value of cultural products, expands domestic product markets, and strengthens the position of the balance of payments. The creative village has become a village with its own special ability to attract tourists compared to other settlement concepts. The types of economic activity offered by Kampung Kreatif involve the local community in their implementation, fostering an atmosphere of sustainable development, and acting as a tool for business development (handicraft producers and small and medium businesses).

4.6.1 Binong Jati Knitted Industry Center

Binong Jati Knitted Industry Center (SIRBJ) is one of the leading industrial centres in Bandung, initiated and run by Binong Jati's community cooperative. It is located on the east side of Greater Bandung, only about 2 km from the city centre. In this community, various kinds of knitted products are produced such as sweaters, cardigans, jackets, scarves, gloves and others (see Figs 4.3 and 4.4). The knitting industry in Binong Jati was first established in the mid-1960s, initiated by five craftsmen who started this business with a contract from a large factory.

What makes SIRBJ relevant to the discussion is its local contribution as one of the small industrial centres that has considerable potential to absorb labour and contribute to the economy of Bandung as part of the tourism thematic. The number of knitting entrepreneurs in SIRBJ increased in 1998 because SIRBJ was able to overcome the crisis that occurred in 1997. In 2015 the

Fig. 4.4. Female labour from the community (the author, 2019).

SIRBJ had a total of 293 units and could absorb 2143 workers. It was ranked second in the number of units and the absorption of labour after the t-shirt manufacturing industrial centres. Moreover, SIRBJ has generated an entrepreneurial community, 'Kampong Radjoet' in which the community connection has initiated the Binong Jati Knitted Industrial Cooperative (KIRBI) where all the members are knitting entrepreneurs in SIRBJ.

KIRBI has become a proactive revival of community-based tourism in favour of 'life chances and welfare' of the marginalized and the less privileged in society. Although the community of Binong Jati is not equipped with adequate systematic tourist industry infrastructure within the neighbourhood (such as accommodation, a tourist information center, entertainment places, cultural institutions, etc), a community of Binong Jati still functions as a host where economic transaction of KIRBI from the knitting products are taking place. Direct benefits from local government interventions in infrastructure can be seen in a renewal programme for roads, small alleys and sign systems to perpetuate not only tourism activity, but also to enhance the daily life of the community. The relationship between tourism and the community can range from one that is wholly in support of tourism to one that has only a minor role in tourism, or even opposes it.

At the beginning of its establishment, KIRBI proceeded very smoothly with a spirit of togetherness and solidity. As a result, SIRBJ knitwear, which controlled 30% of the knitting market in Indonesia, became the largest

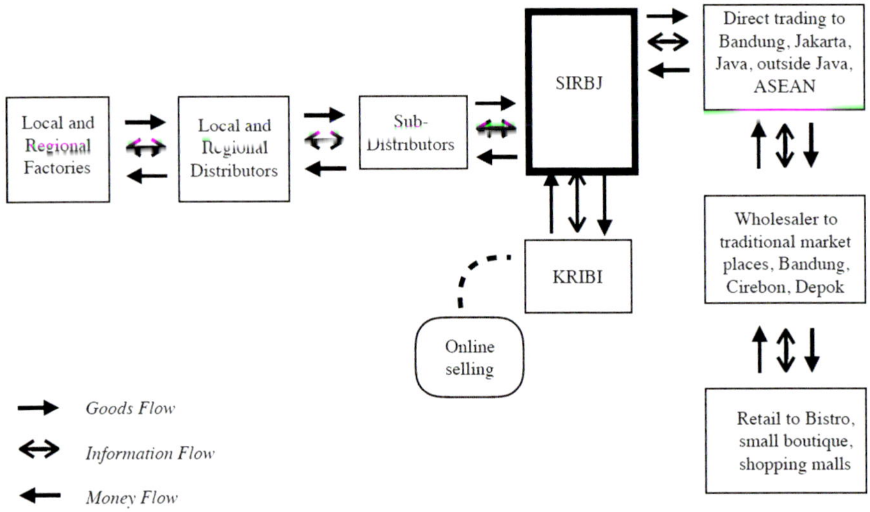

Fig. 4.5. The ecosystem of SIRBJ (data compiled by author, 2019).

distributor in the Tanah Abang (Jakarta) wholesale market, which is one of the largest such markets in the Asian region.

Poor economic conditions and stagnation in the national economic cycle is reflected in lower purchasing power for knitting products, and has led to a decline in the number of knitting entrepreneurs in SIRBJ. In 2008 the next generation of SIRBJ's entrepreneurs created a local initiative with the aim of creating a platform for KIRBI as a place to convey aspirations and information and, especially, to create wider connections and a network of partnerships in marketing to the public through the internet. In 2015, SIRBJ, through the management of KIRBI, experienced a revival due to the entrepreneurial grant programme for creative industry provided by Indonesian President Jokowidodo. After July 2016, the number of entrepreneurs in SIRBJ increased quite significantly. This increase occurred because of the presence of online sales media that caused sales increases. Transaction between individual knitting producers in SIRBJ, the 'Kampoeng Radjoet' community, the KIRBI cooperative (organisation), suppliers and Bandung Municipality implies the existence of a peculiar social-cultural milieu that favours capital accumulation and allows for the combination of resources to set up local business (Fig. 4.5). Added to that, this ecosystem is supported by Kreatif Kampung programmes promoted by Ridwan Kamil, the Mayor of Bandung, started in 2016.

At the start of SIRBJ training, entrepreneurs were dependent on government assistance, but after training was complete and the residents were accustomed to creative activities, each business unit became independent and easily adapted to changes in the economic activities of the city/country. Based on a review of the characteristics of the two communities, it can be seen that a creative community can be formed from various individuals and agencies. However, both communities have clear institutions and coaching contracts so that the creative village can develop until the population is able to carry out tourism activities independently.

The presence of home-based businesses and entrepreneurs in this area implies the existence of a peculiar sociocultural milieu that favours capital accumulation and allows for a combination of resources to set up a business. In addition to selling to the local market, SIRBJ began exporting knitted products to neighbouring countries such as Malaysia and Singapore. In order to be able to promote their products, KIRBI and the Department of Trade and Industry of Bandung, in cooperation with the Department of Culture and Tourism, established a community programme of mentoring and training to develop the entrepreneurial skills of the artisans. The two departments are developing an exhibition centre in the neighbourhood to enable artisans to sell their own products without distributing them to boutiques. In this way, it is believed that distributors' chains will be reduced so that artisans will develop more management skills and attain knowledge on how to promote their own products without depending on boutiques or shopping malls.

During an interview with the chairman of KIRBI, Mr Eka Rahmat Jaya, on 7 January 2019, it was revealed that there are a number of obstacles that interfere with the sustainability of the Kampung Kreatif in Binong Jati, such as: lack of supportive government policy in the case of reducing the price of expensive imported raw materials; SIRBJ entrepreneurs being less ready to compete in comparison to products from China (with the start of the ASEAN Economic Community); management structure not being optimal, for example personal financial reports and business financial reports not being separated, and human resources still limited to family members and surrounding communities in the Binong Jati region; and the underdevelopment of guidance carried out by the Bandung City government in carrying out activities and policies related to small- and medium-sized businesses.

He commented that through KIRBI, the marketing of products is done by online sales and assisting the distribution of products to market centres in Jakarta and Surabaya. Mr Eka is also seeking to establish Binong Jati Kampong as a tourist destination and, a few years ago when he was asked to become head of the village, he said:

> Not all the people in this kampong are open to accepting the concept of kampong tourism, so we introduced socialization through local events. We received support from local municipality, academics, creative community and other entrepreneurial stakeholders to further develop kampong tourism.

Meanwhile, Bandung City Culture and Tourism Agency representative, Mrs Tini Martini Tapran, understands that tourism activity in this community will improve the quality of life in Binong Jati as it is already a well-known knitting centre, but for successful kampong tourism it should be enhanced by accessibility, attractions and amenities. She argued:

> ... tourism must have 3A, accessibility, attractions, and amenities. In this case Kampung Kreatif Binong Jati has good attractions, only lack of amenities; how do people [become aware] that the village can be comfortable, pleasant, calm, neat, and clean?

According to Mrs Tini, the creation of a tourism village in Binong Jati has the additional benefits of adding cultural and educational value by allowing visitors to get to know the community through tours and knitting workshops:

> There must be tour packages offered by the people here, so it's not just shopping, but when they come here, they also get knowledge.

Both Mr Eka and Mrs Tini agree that the community has a vital role in building the destination brand of Kampung Rajut Binong Jati as one of shopping, where daily activities related to knitted products become a positive value that can be sold to tourists. Both the initial activities of producing knitted products (such as preparing raw materials, making patterns, checking machines) and the final stages of production contribute valuable knowledge and education that can be offered to tourists. Through these activities, a harmonious interaction can be formed between tourists and the community, where there are many philosophical values that can also be conveyed.

There are however problems that occur with regard to MSMEs in Bandung, especially the SIRBJ, in terms of decreases in unit numbers and growth. In terms of the programmes carried out by the Bandung City government, there is a gap between elements in this business ecosystem which will potentially make it increasingly difficult for SMEs in Indonesia to survive or compete with foreign products. However there has been a recent increase in independence and level of initiative among the entrepreneurs of SIRBJ. Since entrepreneurial skills also rely on the concept of networking, local networks are highly relevant. The economy of MSMEs is wholly entangled within a specific local culture that regards highly the participation of the family in the work enterprise. From a social perspective, local entrepreneurship seems to be the panacea in sustaining and regenerating small local business enterprise, allowing businesses to locate and organize themselves as producers and sellers.

While writing this chapter the world was struck by the coronavirus pandemic which has had a tremendous impact on tourism. As the phenomenon being studied is ongoing, most of the references come from the mass media and not from scientific articles published in journals, because there are not yet a significant number of papers about the impact of coronavirus on global tourism. Indonesia is one of the countries that is affected by the pandemic and, as of February 2020, the Indonesian government has set aside 443 billion Indonesian rupiah to encourage domestic tourists to visit tourist destinations in Indonesia.

The pandemic has had an impact on many sectoral lines. In the West Bandung Regency, according to data compiled by the West Bandung Creative Economy Forum (FEKRAF KBB), 96.90% of the creative economy sector was affected by COVID-19, including the tourism cluster. However, according to a statement from KBB FEKRAF chairman Mr Teja Sukmana (Nursyabani (2020) in Ayobandung.com, 30 April 2020), creative economic actors who have not yet been affected by the pandemic are those in the fashion sector, which ultimately shifted its production to masks and PPE. On the other hand, the impact on the creative economy sector led to almost 49.25% of workers being laid off. Anticipating this situation, FEKRAF KBB, which is a legal-formal forum for creative economy in West Bandung, conducted fundraising activities to help alleviate the conditions of workers affected by COVID-19. Kampung Rajut Binong Jati could not escape the effects of the pandemic. The impact caused production to plummet by 70%, in part because one of the distribution centres in Tanah Abang, Jakarta, has been temporarily closed and nearly 50% of the workforce have stopped operating.

4.7 Conclusion: Contextualizing the Degrowth Narrative to Community-based Tourism in the Global South

Although the degrowth paradigm in the Global South is not equivalent to that in the Global North, the social innovation solution derived from community initiatives and local resources represents an important core element of degrowth, due to the transformation of the 'consumer oriented' perspective into the 'community oriented'.

Tourism development that considers a 'community oriented' paradigm generates the opportunity for the local community to benefit from the industry. I call it 'community viability', where tourism utilizes the community connection as an active actor for operating tourism amenities and as a further spur to the growth of the community. The workforce has the flexibility to enter and exit the market and many within it often have another, second, job somewhere else. The development of local petty commodity production has a significant role from both the perspective of cultural change and of local economic development. It should be taken into account that it is often the lack of an official licence and withdrawal from the taxation system that classifies an economic activity as informal, not its scale, organization or function (Drakakis-Smith, 1987, p. 68; Evers, 1991, p. 34; Dahles, 2001).

Behind their strategic role in the national and regional economy, SMEs still often face various challenges, both internally, such as limited capital and technology, and externally, in terms of issues such as licensing, raw materials and marketing into regional and global production chains. The ecosystem of local entrepreneurship, can be a formidable 'degrowth from bottom' strategy based on community and an integrated cooperative system between all stakeholders. Furthermore, degrowth in urbanized areas also means resilience of a city related to the ability of its residents to anticipate and participate in overcoming problems that arise.

Infrastructure improvement will not function properly if the individual who uses it is unable to utilize that infrastructure effectively. In order to attract individuals with the right competences and skills, creative coaching is needed for the community. Coaching can be done in various ways, namely self-help by the village community, by the community, or directly by local government. However, to form a sustainable creative village, a clear and integrated system of stakeholders is needed. The coaching process can be distinguished by categories of interest. Supporting actors are tasked with conducting advocacy, coaching, mentoring and monitoring the implementation of creative activities by the community in the village. Key stakeholders are tasked with supporting activities, providing the necessary facilities and infrastructure, and developing funding support.

Tourism in Binong Jati is not only constructed based on its place imagery and identity but also on its utilization of local resources. What can be observed from it is an illustration of how local attributes can be made available as tourist commodities, in a collaboration between tourist actors and stakeholders, providing advantages for the local community. However, the resilience of community-based tourism has been tested during the pandemic, and a lot of craftspeople and local businesses are at risk from the economic impact of the outbreak. Concerns over travel and mobility have hit tourism in general. During the announcement of the first fiscal stimulus package, some of the spending allocations raised questions, such as 72 billion rupiah (equivalent to 4,439,779,20 euro) for social media influencers to promote tourism destinations and a 298.5 billion (equivalent to 18,399,151,95 euro) incentive for airlines and travel agents to attract domestic tourists (Mufti (2020) in thejakartapost.com, 9 April 2020).

Although it is widely acknowledged that tourism and its very nature of commoditization can cause disruptive images and commodity identity, in this extreme situation it is also community capacity and viability that can help in mitigating such a crisis. In this sense, tourism has engaged with elements of degrowth, which is understood within the context of sustainable development and the relationship between tourism and other forms of alternative development.

References

Aldrich, H. (1999) *Organizations Evolving*. Sage Publications, London.

Andriotis, K. (2000) Local community perceptions of tourism as a development tool: the island of Crete. PhD thesis. Bournemouth University, Bournemouth, UK.

Asker, S., Boronyak, L., Naomi, N. and Paddon, M. (2010) Effective community based tourism: a best practice manual. Institute for Sustainable Futures. University of Technology, Sydney. Available at: http://www.isf.uts.edu.au/publications/askeretal2010effectivecbt.pdf (accessed 24 August 2019).

Badan Pusat Statistik kota Bandung (2017) Report, 07.2017 (accessed July 2017, unpublished).

Bandung Spatial Planning Report (2018) August 2018, unpublished.

Bartling, H. (2007) Immigrants, tourists and the metropolitan landscape. In: Church, A. and Coles, T. (eds) *Tourism, Power and Space*. Routledge, Abingdon, pp. 108–122.

Boillat, S., Gerber, J.-F. and Funes-Monzote, F.R. (2012) What economic democracy for degrowth? Some comments on the contribution of socialist models and Cuban agroecology. *Journal Futures* 44, 600–607.

Butler, S.R. (2012) Curatorial interventions in township tours: two trajectories. In: Frenzel, F., Koens, K. and Steinbrink, M. (eds) *Slum tourism. Poverty, power and ethics*. New York, pp. 215–231.

Castells, M. (1996) *The Rise of the Network Society*. Blackwell, Oxford.

Cukier, J. (1996) Tourism employment in Bali: trends and implications. In: Butler, R. and Hinch, T. (eds) *Tourism and Indigenous Peoples*. International Thomson Business Press, London, pp. 49–75.
Dahles, H. (1997) Tourism, petty entrepreneurs and sustainable development. In: Dahles, H. (ed.) *Tourism, Small Entrepreneurs and Sustainable Development: Cases from Developing Countries*. ATLAS/Tilburg University, pp. 23–33.
Dahles, H. (2001) *Tourism, Heritage and National Culture in Java. Dilemmas of a Local Community*. Curzon, London.
De Kadt, E. (ed.) (1979) *Tourism: Passport to development?* Oxford University Press, New York.
Degrowth Leipzig Conference (2014) What is degrowth? Available at: https://www.degrowth.info/en/what-is-degrowth/ (accessed 24 August 2019).
Demaria, F., Schneider, F., Sekulova, F. and Martinez-Alier, J. (2013) What is Degrowth? From an activist slogan to a social movement. *Environmental Values* 22(2), 191–215. DOI: 10.3197/096327113X13581561725194.
Denzin, N.K. and Lincoln, Y.S. (1994) *Handbook of Qualitative Research*. Sage Production, Inc, USA.
Department Pariwisata dan Budaya Bandung (2017) 09.2017.
Department Perdagangan dan Perindustrian Bandung Report (2017) 09.2017.
Drakakis-Smith, D.W. (1987) *The Third World Cities*. Routledge, Routledge Introduction to Development. Abingdon, UK.
Escobar, A. (2015) Degrowth, post development, and transitions: a preliminary conversation. *Sustainability Science* 10(3), 451–462.
Evers, H.D. (1991) Shadow economy, subsistence production and informal sector. Economic activity outside of market and state. *Prisma* 51, 34–45.
Fainstein, S.S. and Judd, D.R. (1999) Global forces, local strategies, and urban tourism. In: Judd, D.R. and Fainstein, S.S. (eds) *The Tourist City*. Yale University Press, New Haven.
Firman, T. (2002) Urban development in Indonesia, 1990-2001: from the boom to the early reform era through the crisis. *Habitat International* 26, 229–249.
Freire-Medeiros, B. (2009) The favela and its touristic transits. *Geoforum* 40(4), 580–588. DOI: 10.1016/j.geoforum.2008.10.007.
Geertz, C. (1965) *The Social History of an Indonesian Town*. Greenwood Press, Westport, Connecticut.
Ghezzi, S. (2012) Erasga, D. (ed.) *Small-Scale Entrepreneurship in Modern Italy - An Ethnographic Analysis of Social Embeddedness in the Access to Capital and Credit, Sociological Landscape - Theories, Realities and Trends* ISBN: 978-953-51-0460-5.
Gotham, K.F. (2005) Tourism Gentrification: the case of new Orleans' Vieux Carre (French quarter). *Urban Studies* 42(7), 1099–1121. DOI: 10.1080/00420980500120881.
Gotham, K.F. (2007) *Authentic New Orleans: Tourism, culture and race in the big easy*. New York University Press.
Guinness, P. (1989) Social harmony as ideology and practice in Javanese city. In: Alexander, Paul. (ed.) *Creating Indonesian Cultures*. University of Sydney, Oceania Publications, Sydney, pp. 55–74.
Harvey, D. (1989) *The Urban Experience*. Johns Hopkins University Press, Baltimore.
Harvey, D. (2005) *A Brief History of Neoliberalism*. Oxford University Press, Oxford, UK.
Hugo, G.J. (1996) Urbanization in Indonesia: City and countryside linked. In: Gugler, J. (ed.) *The Urban Transformation in the Developing World*. Oxford University Press, Oxford, pp. 133–183.
Husbands, W. (1981) Centers, peripheries, tourism and socio-spatial development. *Ontario Geography* 17, 37–59.

Jafari, J. (1989) Sociocultural dimension of tourism: an English language literature review. In: Bustrzanowski, J. (ed.) *Tourism as a Factor of Change: a Sociocultural Study*. Economic Coordination Centre for Research and Documentation in Social Science, Vienna, pp. 17–60.

Jamal, T.B. and Getz, D. (1995) Collaboration theory and community tourism planning. *Annals of Tourism Research* 22, 186–204.

Jones, G.W. (2002) Southeast Asian urbanization and the growth of Mega-Urban regions. *Journal of Population Research* 19(2), 119–136.

Kallis, G., Kerschner, C. and Martinez-Alier, J. (2012) The economics of degrowth. *Ecological Economics* 84, 172–180. DOI: 10.1016/j.ecolecon.2012.08.017.

Kallis, G., Kostakis, V., Lange, S., Muraca, B., Paulson, S. *et al.* (2018) Research on Degrowth. *Annual Review of Environment and Resources* 43(1), 291–316. DOI: 10.1146/annurev-environ-102017-025941.

Kim, P.H. and Aldrich, H. (2005) *Social Capital and Entrepreneurship*. Now Publishers Inc., Hanover, MA.

Latouche, S. (2009) *Farewell to Growth*. Polity, Cambridge, UK.

Martínez-Alier, J. (2012) Environmental justice and economic degrowth: an alliance between two movements. *Capitalism Nature Socialism* 23(1), 51–73. DOI: 10.1080/10455752.2011.648839.

Martínez-Alier, J., Pascual, U., Vivien, F.-D. and Zaccai, E. (2010) Sustainable de-growth: mapping the context, criticisms and future prospects of an emergent paradigm. *Ecological Economics* 69(9), 1741–1747. DOI: 10.1016/j.ecolecon.2010.04.017.

Mathieson, A. and Wall, G. (1982) *Tourism: Economic, Physical and Social Impacts*. Longman, London.

McCannell, D. (1992) *Empty Meeting Grounds: The Tourist Papers*. Routledge, London.

McGee, T., Lin, G.C.S., Wang, M., Marton, A. and Wu, J. (2007) *China's Urban Space: Development Under Market Socialism*. Routledge Studies on China in Transition. Routledge, Abingdon, UK.

McIntosh, R.W. and Goeldner, C.R. (1986) *Tourism: Principles, Practices, Philosophies*, 5th edn. Wiley, New York.

Mowforth, M. and Munt, I. (eds) (2009) *Tourism and Sustainability – Development, Globalisation and New Tourism in the Third World*, 3rd edn. Routledge, Abingdon, UK.

Mufti, R.R. (2020) Tourism sector suffers 1.5 billion losses as Covid-19 hits hard. Available at: https://www.thejakartapost.com/news/2020/03/16/tourism-sector-suffers-1-5-billion-losses-as-covid-19-hits-hard.html (accessed 9 April 2020).

Murti, K. (2015) Babad Kampung" Celebrating History and Neighbourhood Identity in Yogyakarta. In: Hatley, B. and Hough, B. (eds) *Performing Contemporary Indonesia: Celebrating Identity, Constructing Community*. Koninklijke Brill NV, Leiden, The Netherlands.

Neuman, W.L. (2000) *Social Research Methods: Qualitative and Quantitative Approaches*, 4th edn. Allyn and Bacon, Boston.

Nursyabani, F. (2020) 96,90% Sektor Ekonomi Kreatif di Bandung Barat terdampak Covid-19. Available at: https://ayobandung.com/read/2020/04/30/87709/9690-sektor-ekonomi-kreatif-di-bandung-barat-terdampak-covid-19 (accessed 30 April 2020).

Page, S.J. and Hall, C.M. (2003) *Managing Urban Tourism*. Harlow, Prentice Hall.

Podlaszewska, R.S.A. (2017) Placemaking of Destination. Localization and Commoditization Process of Urban Tourism. Dissertation Thesis. Bauhaus-Universität Weimar. Available at: http://nbn-resolving.de/urn:nbn:de:wim2-20171201-36995

Richter, L.K. (1992) Political instability and tourism in the third world. In: Harrison, D. (ed.) *Tourism in the Less Developed Nation*. Belhaven, London, pp. 35–46.

Rodríguez-Labajos, B., Yánez, I., Bond, P., Greyl, L., Munguti, S. *et al.* (2019) Not so natural an alliance? Degrowth and environmental justice movements in the global south. *Ecological Economics* 157, 175–184. DOI: 10.1016/j.ecolecon.2018.11.007.

Rothman, J. (1968) Three models of community organization practice. In: Cox, F., Erlich, J.L., Rothman, J. and Tropman, J.E. (eds) *Strategies of Community Organization*. F.E. Peacock Publishers, Inc, Itasca, Illinois, pp. 20–36.

Sassen, S. (1991) *The Global City*. Princeton University Press, Princeton.

Selby, M. (2004) Consuming the city: conceptualizing and researching urban tourist knowledge. *Tourism Geographies* 6(2), 186–207. DOI: 10.1080/1461668042000208426.

Shaw, G. and William, A.M. (1990) Tourism economic development and the role of entrepreneurial activity. In: Cooper, C.P. (ed.) *Progress in Tourism*. 2. Recreation and Hospitality Management, Bellhaven, London, pp. 67–81.

Shaw, G. and William, A.M. (1998) Entrepreneurship, small business culture and tourism development. In: Ioannides, D. and Debbage, K.G. (eds) *The Economic Geography of the Tourist Industry*. Routledge, London, pp. 235–255.

Shaw, G. and Williams, A.M. (1994) *Critical Issue in Tourism: A Geographical Perspective*. Blackwell Publishers, Oxford.

Sihombing, A. (2004) *The Transformation of Kampung Kota: Symbiosis Between Kampong and Kota, a Case Study from Jakarta*. University of Indonesia, Department of Architecture.

Swedberg, R. (2000) *Entrepreneurship: The Social Science View*. Oxford University Press, Oxford.

Urry, J. (1995) *Consuming Places*. Routledge, London.

Weiss, M. and Cattaneo, C. (2017) Degrowth - Taking stock and reviewing an emerging academic paradigm. *Ecological Economics* 137, 220–230. DOI: 10.1016/j.ecolecon.2017.01.014.

Willmott, P. and Young, M. (1960) *Family and Class in a London Suburb*. Routledge and Kegan Paul, London.

Young, G. (1973) *Tourism - Blessing or Blight?* Penguin, London.

5 Political Discourse Analysis of the Degrowth Challenge to Dominant Tourism Narratives in Spain

Neil Hughes[1*] and José Mansilla[2]

[1]*University of Nottingham, Nottingham, UK*
[2]*University of Lleida, Barcelona, Spain*

5.1 Introduction

The sight of anti-tourism protest is increasingly common on the streets of major tourism destinations across the world (see case studies in Colomb and Novy, 2016). In cities as diverse as Amsterdam, Barcelona, Bilbao, Berlin, Cartagena de Indias in Colombia, Dubrovnik, Florence, Hvar, Milan, Palma and Venice, local inhabitants have mobilized to demonstrate their anger at the detrimental effect tourism is having on both their lives and on the environment in which they live. This is particularly prevalent in contexts such as Barcelona (e.g. Hughes, 2018) in which a significant increase in tourism has been witnessed due to public authorities' efforts to manage the transition from an industrial past to a present development model in which cultural capital and geographical location are important assets. The extent of the discontent is reflected in the in-depth treatment anti-tourism contestation receives in media outlets as diverse as the *New Yorker* (Mead, 2019), *Time* (Abend, 2018), *The Conversation* (Milano *et al.*, 2018) and the BBC (Campoamor, 2017). As Higgins-Desbiolles *et al.* (2019) point out in their analysis of degrowth ideas and tourism, at the heart of these mobilizations lies a critique of an elite discourse that extols tourism's positive impact on the economic, political, social and cultural life of the destinations in which it takes place. No more is this the case than in Spain, where protestors have poured scorn on official claims praising tourism's virtues.

To explain the dominant discourse and the challenge degrowth poses to it in Spain, this chapter sets out a conceptual framework that draws on ideas

*Corresponding author: neil.hughes@nottingham.ac.uk

DOI: 10.1079/9781789245073.0005

from political discourse analysis (PDA), treatments of both ideology and interpellation (Althusser, 1970) and the wider work of Higgins-Desbiolles *et al.* (2019) on degrowth and tourism. It argues that by emphasizing the contribution tourism has made to both economic and political development for decades in Spain, dominant tourism discourse provides a positive frame for the industry that is in sharp contrast to the concerns about its harmful effects articulated by its opponents. It shows, using examples from speeches, articles and reports delivered by government ministers and tourism authorities, how official discourse interpellates Spanish citizens as convivial subjects with responsibility for creating a hospitable environment in which international tourists can feel both welcome and safe. The chapter also demonstrates what happens when such attempts at interpellation misfire, as has happened so spectacularly in the case of Barcelona and other important Spanish tourism destinations in recent years. In its analysis of the Barcelona case, it identifies the social actors leading the campaign against the dominant discourse and the vested interests it protects as a coalition of civil society groups, organizations and assemblies coordinated increasingly by the degrowth-inspired *Assemblea de Barris per un Tourisme Sostenible* (Neighbourhood Assembly for Sustainable Tourism, henceforth referred to as the *ABTS*). This association, it argues, has developed a powerful counter-narrative and waged an effective campaign against what it sees as a neoliberal tourism project that constitutes an existential threat to social cohesion and environmental sustainability in the city. Finally, it looks at the alternative policy proposals the *ABTS* advocates and assesses the extent to which they have impacted tourism decision-making in Barcelona.

5.2 Theoretical Framework

PDA's leading proponents such as van Dijk (1997) identify the object of study of this growing sub-field of critical discourse analysis (CDA) as the text and talk produced by political actors, primarily, but not exclusively, during political events or practices ranging from election campaigns to televised debates and from party conferences to appearances before parliamentary committees. By focusing on discursive aspects of politics, PDA shines a novel analytical light on important issues in political science such as the discursive production and reproduction of political power and dominance within society and of the various forms of resistance to it from non-governmental organizations, pressure groups and protest movements such as the ones discussed in this chapter (van Dijk, 1997, p. 13).

The detailed analysis of the functions of political texts, the intentions embodied within them and their social and political implications, requires a methodological approach that combines the study of both language and context. Regarding context, van Dijk (1997, pp. 16–18) identifies a range of important contextual factors that interact with text to create meaning. These include political systems, institutional arenas, political parties, features of the wider non-state actor community, and ideology, which is seen as a system of representation that serves to produce (and reproduce) a common-sense consensus around the ahistoricity and immutability of prevailing economic, political and social relations. By defining such relations as, at worst, benign, ideology seeks to remove them from the realm of contestation. As Freeden (1996) argues:

> an ideology attempts to end the inevitable contention over concepts by decontesting them, by removing their meanings from contest … [ideology] is a wide-ranging structural arrangement that attributes decontested meanings to a range of mutually defining political concepts.

For most political discourse analysts, ideologies play a key role in shaping individual and group consciousness about their social role, rank, function and influence. Through his development of the concept of interpellation, Althusser (1970) explains how such subjectivities are fixed via a process of 'hailing', whereby individuals are assigned subject positions in the social formation they inhabit:

> I shall then suggest that ideology 'acts' or 'functions' in such a way that it 'recruits' subjects among the individuals (it recruits them all), or 'transforms' the individuals into subjects (it transforms them all) by that very precise operation which I have called interpellation or hailing. (Althusser, 1970, pp. 173–174)

Although it accepts the theoretical usefulness of the concept of interpellation, this chapter is cognizant of one of the main criticisms of Althusser's work i.e. that it is overly deterministic and limits the scope for agency of non-hegemonic groups and actors (e.g. Berger and de Cleen, 2018, pp. 94–95). Consequently, the chapter looks at what happens when interpellation misfires and individuals respond not to the official discourse, but to the call of political actors with an alternative narrative that represents a challenge to the hegemonic status quo.

As will become apparent in the discussion of tourism, rhetorical function and the idea that language is a form of action is a key issue in PDA. These important discourse features are coded in expression and can be revealed through careful linguistic enquiry. The properties of language considered by PDA scholars such as van Dijk (1997) include sentence types (imperatives, interrogatives or exclamations), text genres (e.g. speeches and political slogans), aspects of semantics (tense, lexis, syntax, etc.), phonological categories (e.g. intonation) and the range of rhetorical forms and devices traditionally associated with political speech. In the case of semantics, for example, pronominal polarization is often used in political discourse to divide the world into 'us' and 'them' positions. Such Manichean rhetoric, we will argue, can be detected in the language deployed by degrowth activists in slogans such as 'tourist you are the terrorist' that have been used by activists linked to the left-nationalist *Candidatura d'Unita Popular* (*CUP*) political party to 'other' tourists as inimical to the public interest. In can also be seen in elite attempts to denounce such activism as the work of a 'tourismphobic' outgroup whose actions stand in sharp contrast to the 'friendliness' and 'conviviality' displayed by the rest of Spanish society.

The final constitutive element of the multidisciplinary theoretical context framing this chapter is provided by the contribution of Higgins-Desbiolles *et al.* (2019) to the degrowth and tourism debate. In their work, they assess tourism's contribution to environmental destruction and climate change and reflect on the relationship between other degrowth priorities such as tourism's impact on social justice and questions of accountability in tourism decision-making. They also set out some of the concrete policies advocated by the proponents of degrowth to address their concerns. These include redefining tourism in ways that prioritize the needs, interests and participation of local communities; a shift away from the deification of growth towards a model based on principles such as '*buen vivir*', which according to Higgins-Desbiolles *et al.* (2019, p. 12) implies abandoning 'measures of prosperity based on gross domestic product for measures of people's' total well-being.'; an increased role for regulation 'for sustainability and the public good'; the rejection of neoliberal discourses that see regulation as an obstacle to economic competitiveness; and a shift away from tourism dependency towards an economic model based on greater diversity. One of the central aims here is to explore to what extent this panoply of concerns and recommendations has become diffused into the Spanish context.

5.3 The Ideological Discourse of Tourism in Spain

According to Freeden (1996), one way of thinking about ideological discourses is as conceptual maps that help us navigate the contours of social reality. In this section, we identify the discourses and themes that constitute the 'map' that guides us through the social reality of tourism and establishes subject positions in relation to this most important of Spanish industries. This is important given the claim by Higgins-Desbiolles *et al.* (2019) that degrowth constitutes, above all, a challenge to a discourse that sees tourism as an overwhelmingly positive asset for destinations and their inhabitants. It is also necessary due to the emphasis that degrowth activists in Spain themselves place on what they see as a 'battle of ideas' against an 'official, public–private story' that depicts tourism as 'natural and unquestionably positive' (ABTS, 2019).

The key coordinates on the ideological map of Spanish tourism include notions of tourism as a positive agent for economic, political and social change in the country; a 'goose that lays the golden egg' narrative that emphasizes the need to avoid practices that might be considered antagonistic or inimical to its reproduction; a discourse of national pride and unity evident at both the national and regional levels; claims around the role of business leadership in driving tourism growth; an increasingly prevalent sustainability narrative; and what will be referred to as a 'discourse of conviviality' that hails Spanish citizens to a subject position requiring the exercise of hospitality and the creation of a welcoming and safe environment. As will be demonstrated, such discourse is embedded across a body of texts that include speeches, interviews, articles, legislation and reports produced by elite political and social actors at the national, regional and municipal levels in Spain.

5.4 'Tourism for a Better Spain'

In April 2019, Isabel Oliver Sagreras marked her appointment as Spain's Secretary of State for Tourism with an article in the leading national daily, *El País*, under the title, 'Tourism for a Better Spain' (Oliver Sagreras, 2019). In it, the new secretary of state demonstrated her mastery of the dominant tourism discourse's key themes, particularly the role that tourism has and continues to play in Spain's development, describing it as a:

> formidable agent of social and economic transformation ... that has made us into the prosperous, modern, tolerant, hospitable and safe country of which we should all feel proud. We will use our extraordinary capacity to make others happy in order to build bridges and construct together this better Spain that we all deserve. (Oliver Sagreras, 2019)

In her piece, Oliver Sagreras draws on important historical references such as the transition to democracy in the 1970s following the end of dictatorship in Spain, to construct an idea of tourism that resonates positively with the Spanish public:

> Tourism is one of the main driving forces in the world, which has changed authoritarian regimes, has created employment in depressed areas, has motivated the construction of vital infrastructure and has helped to establish peace in post-conflict areas. Without going any further, in Spain's dictatorship, tourism was an open window to other ways of living, a catalyst that facilitated the transition of society towards democracy. (Oliver Sagreras, 2019)

Such discourse about tourism's positive contribution to Spain's political development is also common in the abundant academic literature about the Spanish transition to democracy. In it, tourism is identified as a key factor in the cultural change that proceeded the political transformation experienced in the 1970s. In a typical example of this thinking, Sánchez Sánchez (2001, p. 220) argues that 'direct personal encounters between tourists and residents facilitated the Spanish population educated under Franco access to the patterns of social, moral and cultural behaviour common in neighbouring countries'.

The idea that tourism should be seen as the 'goose that lays the golden egg' is also reflected in Oliver Sagreras' claims around tourism's economic impacts:

> This sector is absolutely strategic for Spain. It compensates our balance of payments, employs hundreds of thousands of people, generates wealth in the territory and creates companies. We have been building the most competitive country in the world for 50 years, according to the World Economic Forum. We are the second largest destination in the world in revenue (about 90,000 million euros in 2018) and in international tourists (82.8 million). (Oliver Sagreras, 2019)

More recently, the 'golden egg' narrative has also been present in the criticism directed at those deemed to be placing the industry's competitiveness at risk due to their unfounded claims about the negative consequences of tourism. In Barcelona, for example, where tourism degrowth contestation has been to the fore, the business community and its advocates in the press have been particularly critical of what it describes as 'tourismphobic' attitudes that are pitching the local community against tourists. It can also be seen in ministerial discourse at the national level. For example, according to Spain's former Secretary of State for Tourism, Matilde Asían, 'tourismphobia could dent Spain's tourism potential ... and we are not going to kill the goose with the golden eggs'. In her efforts to reassure business, Asián signalled the state's willingness to move to a more repressive response if the dominant discourse failed to achieve its interpellatory function: 'whilst the best thing would be to stop such actions, where they do take place, they should be investigated and where there is evidence of criminality [the perpetrators] punished' (*La Nueva España*, 2017).

According to Milano *et al.* (2019), the use of the neologism 'tourismphobia' is part of a concerted effort by the media and other elite actors to divert attention away from legitimate social movement concerns. While bitterly refuted by many within the tourism degrowth movement, their cause has not always been helped by some of the actions of anti-tourism activists who through their use of slogans and graffiti have effectively 'othered' foreign tourists as an existential threat to the city. While '*el turismo mata la ciudad*' ('tourism kills the city') is arguably the most common slogan in the genre, many others, in Catalan, English and Spanish have disfigured the city's elegant facades such as: 'Why call it tourist season if we can't kill them?'; 'all tourists are bastards'; '*hipsters y turismo, nuevas formas de terrorismo*' ('hipsters and tourism, new forms of terrorism'); '*Barcelona no esta en venda*' ('Barcelona isn't for sale'), 'tourist go home or die' and 'tourist you are the terrorist'.

As Hughes (2018) discusses, such slogans have functioned as a 'call to arms' to young activists to take action against tourists to protect the city and return it to its people. In August 2017, for example, a group of young activists were reported to have stopped a sightseeing bus close to the *Camp Nou* football stadium, slashed its tyres and painted the slogan, 'tourism kills neighbourhoods' on its side. Videos of the attack were uploaded to YouTube shortly after the event, accompanied by the caption, 'mass tourism kills neighbourhoods, destroys the territory and condemns the working class to misery'. In other incidents, for-hire bicycles had their tyres punctured and several hotels and restaurants were paint-bombed. Again, the videos' accompanying slogans appear to legitimize the vandalism as do subsequent justifications for the attacks issued by the activists in which they describe their actions as acts of 'self-defence' against the tourism invasion.

Dominant tourism discourse attributes the sector's perceived success to the Spanish business community, which, it is claimed, through a combination of hard work and acumen, has turned Spain into one of the world's leading tourism destinations. According to the right of centre newcomer to the increasingly plural political party scene in Spain, Contigo Somos Democracia (2019):

> If Spain today is one of the world's main tourist destinations, it is solely and exclusively because of the magnificent work of the business sector ... Entrepreneurs and workers, together with the excellent weather, beaches, the sea, culture, traditions and gastronomy, among others, are really the ones responsible for our prestige and international tourist recognition.

The fears and anxiety within the tourism industry about the effects of anti-tourism activism have led to the emergence of a discourse that places responsibility for creating a warm, friendly and hospitable environment squarely on the shoulders of the Spanish people. This discourse, which we will refer to as 'a discourse of conviviality', can be detected at all institutional levels in Spain. At the national level, for example, Oliver Sagreras unambiguously sets out, in her article for *El País* (Oliver Sagreras, 2019), Spanish citizens' responsibility for ensuring tourists' happiness and safety while on holiday in Spain:

> Let's continue using tourism as the formidable agent of social and economic transformation that has always been for Spain, and that has made us the prosperous, modern, tolerant, hospitable and safe country that we can and should be proud of. Let's use our extraordinary ability to make others happy to build bridges and build that better Spain that we all deserve.

In Barcelona, *Turisme de Barcelona*, a public/private partnership set up in the wake of the 1992 Barcelona Olympics with representation from Barcelona Municipal Council, the Barcelona Official Chamber of Commerce, Industry and Shipping and the Barcelona Promotion Foundation, described tourism as 'one of our crown jewels' that creates wealth and employment in the city. According to the consortium, it is incumbent on Catalans to behave appropriately towards tourists if the city is to retain its attractiveness as one of Europe's most important city destinations:

> Barcelona is and must remain an attractive destination internationally for its quality and conviviality. It is a city where tourism supports the daily life of its inhabitants and we are proud to be people of Barcelona. (Cited by Zerva *et al.*, 2019, p. 318)

In one of the most unerring examples of the discourse of conviviality, as part of a 2018 publicity campaign by the Valencian Tourism Agency, the local community was urged to 'spread its hospitality'. According to the Agency, 'welcoming and hospitality' are 'part of the Mediterranean character' that 'defines the citizens of the Valencian Community'. The importance of hospitality is linked in the campaign to Valencia's competitiveness as an international tourist destination within the context of wider globalization: 'being more hospitable positions us more and better in a global world'. The campaign is also framed as an antidote to what the director general of tourism sees as an increasingly hostile tourism environment characterized by anti-tourism sentiment: 'The Spread Your Hospitality campaign stands up for tourismphilia against tourismphobia' (Europapress, 2018).

Sector concerns about the impact that anti-tourism discourse is having has spawned a new element in the dominant discourse around tourism, i.e. a sustainability narrative that sees the future of Spanish tourism as tied to a growth model that considers both the industry's environmental and social impacts:

> The Ministry of Tourism is transforming the tourism growth model towards a more sustainable and sustained model. To promote this shift towards sustainability and for the sector to contribute to achieving the United Nations Sustainable Development Goals (SDGs), we have signed a collaboration agreement with the High Commissioner of the 2030 Agenda with the commitment to disseminate the objectives and goals in its training programs, in the Network of Intelligent Tourist Destinations, in its cooperation actions with third countries, in the Spanish Tourism Quality System and in the promotional activities of Turespaña. (Oliver Sagreras, 2019)

This discourse, while at first glance progressive, achieves several key ideological objectives. It positions the state in the guise of the Ministry of Tourism as the leader of initiatives to address questions of sustainability; it simplifies solutions to climate change away from the complex and difficult decisions that a sincere commitment to sustainability might entail, replacing them instead with an international agreement that, to all intents and purposes, is unenforceable by law; and in a claim that is contested by degrowth activists, it suggests that Spain is taking its international responsibilities in this area seriously. Finally, there is a suggestion in the discourse that progress towards sustainability is being made and that tourism, rather than being part of the problem, is making a crucial contribution to attaining this goal:

> It is now, installed in success, when we must reorient our growth model towards the sustainability paradigm embodied in the SDGs of the 2030 Agenda, turning tourism into an instrument in favor of equality, inclusion, decent work, conservation of ecosystems and heritage, depopulation or social cohesion. (Oliver Sagreras, 2019)

5.5 Tourism Activism and Degrowth

The attempt to interpellate agents of conviviality through campaigns such as 'Spread your Hospitality' has come under considerable pressure in recent years. Across the country, a coalition of activists has transformed into a powerful social movement that opposes many aspects of the dominant discourse. By drawing on numerous studies, many of which have an ethnographic focus, the following section maps the contours of this coalition and assesses the extent to which degrowth ideas and activists have been prominent within its organization. This section focuses mainly on Barcelona, the city in which degrowth principles have exercised most influence over the performance of anti-tourism activism in Spain.

Academic interest in tourism contestation and activism in Spain, while still in its infancy, has grown in recent years, particularly among Spanish academics such as Degen (2004), Mansilla (2018a, b) Milano *et al.* (2018) and Milano and Mansilla (2019). This growing body of work has revealed important insights into the geographic locus of activism, the network of actors involved in contesting tourism, the contexts within which they operate, the world views shaping their response to mass tourism, the action repertoires they employ and the impact of their activism on public opinion.

It is clear from the extant literature that the emergence of social movements focused primarily on the pernicious social, economic, cultural and environmental effects of tourism is a relatively recent phenomenon in Spain that has its epicentre in Barcelona. In their work on tourism gentrification in Barcelona, Degen (2004) and Mansilla and Milano (2018) identify signs of coordinated action against the growing impact of tourism on access to public services, spaces and buildings in the early 2000s as the historical point of departure for the analysis of such movements. Milano *et al.* (2018) also emphasize the importance of public concerns around housing, particularly the transformation of 'residential apartments' into 'tourist accommodation' as a significant cause for popular concern at this point in the movement's evolution as well as a culture clash between tourists and locals, 'about the different lifestyles of people who come for tourism purposes and people who have daily-routines'. For Mansilla (2018a, b), ambitious city redevelopment initiatives such as the Plan 22@ that is transforming theformerly industrial Poblenou into the city's technology and innovationdistrict, mega-events from the Olympic Games in 1992 to the Universal Forum of Cultures in 2004, and the 2011 '*indignados*' mobilization have been pivotal factors in the construction of tourism contestation in Barcelona. In the case of the Universal Forum of Cultures, for example, this collaboration between Barcelona City Council, the Catalan Autonomous Government, the Spanish Government and Unesco is one of a number of large-scale international events that have taken place in Barcelona to attract new investment and tourists to the city. The main criticisms of the event focus on the project's environmental impact and its contribution to a wider process of city branding aimed at selling Barcelona to the world as '*la botiga més gran del mon*' (the biggest shop in the world).

The focus on the impact of overtourism on access to public services and goods such as housing as well as the antisocial behaviour of tourists in residential areas was also to the fore in one of the most significant outbreaks of anti-tourism protest – the so-called 2014 'Barceloneta Crisis'. The crisis was triggered when hundreds of locals took to Barceloneta's streets and squares to protest against tourists' uncivilized behaviour, which, they argued, was blighting their neighbourhood (Hughes, 2018; Mansilla, 2018a, b; Mansilla and Milano, 2018). The catalyst for this outbreak of popular anger was an incident in which a group of young tourists was seen to be wandering naked around the streets of a traditionally working-class area while uploading videos of their exploits to YouTube. The protests, which were widely reported both inside and outside Spain, led to the setting up of neighbourhood protest groups such as *La Barceloneta Diu Prou* (Barceloneta Says Enough), *Barceloneta No Es Ven* (Barceloneta is Not for Sale) and *EnsPlantem* (We Stand

Firm) that coordinated subsequent campaigns against the tourist apartment phenomenon and related uncivilized tourist behaviour. The events of 2014 provided the impetus for efforts to unite the disparate anti-tourism struggles that, by this juncture, had broken out in neighbourhoods across Barcelona, into a common front against the economic, social and environmental costs of a developmental model based on mass tourism. Such efforts led to the setting up of the *ABTS*, which, since its creation in 2015, has been widely seen as the main actor in tourism contestation across the city (Hughes, 2018, p. 472; Milano *et al.*, 2018).

The summer of 2017 witnessed a landmark shift in the evolution of the anti-tourism movement. It marks the point at which the phenomenon gained global media attention due, above all, to incidents such as the *Camp Nou* tourist bus attack. The extent of business concern with the turn of events led representatives to urge the Barcelona authorities to take action to curb such excesses. Following the bus attack in August, for example, 27 business associations wrote to the mayor, Ada Colau, to express their 'growing concerns and sense of alarm' and urge her to 'cut out at the root the wave of attacks targeting tourism', which was putting the 'international image of the city in danger' (Hosteltur, 2017). According to Ordíz (2017), the level of popular anger at the 'tourist invasion' had reached such a head of steam by this point that it was becoming 'very complicated to control'.

The other major consequence of the 2017 events in Barcelona was the important imitation effect they had beyond the region, with copycat action being reported in other Spanish cities in which tourism makes a significant contribution to the local economy (Baran, 2017). A cursory search online reveals protests and acts of vandalism against tourism businesses taking place in cities as diverse as Seville, Cadiz, Palma, Malaga, Bilbao, San Sebastian, Madrid, Valencia and on the islands of Tenerife and Ibiza. In the Basque Country, for example, in Bilbao and Sebastian, much of the action was led, as in Barcelona, by young radicals linked to a left-nationalist party, Sortu. In the summer of 2017, such activists were responsible for a wave of anti-tourism incidents including acts of vandalism directed at tourism businesses and services. These culminated in a march through the streets of San Sebastian behind banners claiming, 'this isn't tourismphobia, this is class struggle' (Gorospe, 2017). As part of their response, both businesses and the public authorities were quick to invoke the discourse of conviviality and to impress upon locals the importance of extending hospitality to visitors. In the Basque Country, for example, a campaign titled '*Ongi Etorri*' ('Welcome in Basque') was launched, which according to the Basque Minister for Tourism, Alfredo Retortillo, 'did not aim to increase the number of tourists'; instead the goal was to highlight to visitors that the Basque Country was 'open and welcoming' (Izarra, 2018). In Seville, another city hit by tourism contestation, a local business association (The Association of City Centre Traders) handed out 40,000 free fans to tourists printed with the message, 'Seville loves Tourists, Welcome to your Home' (Ybarra, 2017).

Assemblea de Barris per un Tourisme Sostenible (Neighborhood Assembly for Sustainable Tourism)

The Barcelona-based *ABTS* is the main degrowth-inspired organization focusing on the pernicious social and environmental impacts of tourism in Spain. It integrates more than 30 neighbourhood groups, assemblies and movements that as Hughes (2018, p. 473) points out:

> are united by their rejection of what they see as the neoliberal 'touristification' of their neighbourhoods and commitment to practices and values such as assemblyism and consensus-based decision-making, environmental responsibility, degrowth, and the 'right to the city'.

Many of those involved in tourism degrowth contestation are also political party activists, most commonly in either *Barcelona En Comú* (Barcelona in Common), a citizens' platform that currently governs Barcelona City Council, or the more explicitly anti-capitalist *Candidatura d'Unita Popular, CUP* (Popular Unity Candidacy).

The emphasis in *ABTS*'s discourse on the link between neoliberal restructuring, overtourism and the need for degrowth is articulated by *ABTS* activists in their contribution to the academic literature that has emerged looking at these issues in recent years. In his collaboration with the academic Agustín Cocola-Grant, for example, the ABTS activist Daniel Pardo (Cocola-Gant and Pardo, 2017) identifies neoliberal measures introduced both prior to and, in particular, in the wake of the 2008 financial crisis as one of the main reasons for the spectacular increase in tourism stays in Barcelona, which they estimate at more than 30 million a year. Such measures include the relaxation of urban planning regulations limiting hotel and holiday accommodation in the historical city centre; measures to incentivize tourism businesses such as tax breaks and the loosening of licensing restrictions on businesses; air transport deregulation; labour market reforms; and a lax regulatory environment for tourism accommodation platforms such as Airbnb, all of which, they argue, have contributed significantly to Barcelona's 'touristification'.

Since its inception, ABTS (2019) has followed three 'lines of action' in its challenge to tourism's public and private sector coalition leaders and their claim that tourism growth has been decoupled from its harmful environmental and social effects and turned into an 'instrument' of 'inclusivity', 'social cohesion' and the 'conservation of ecosystems'. As already mentioned, its first line of action has taken the form of a 'battle of ideas' against what the ABTS (2019) describes as an 'official, public–private story' that depicts tourism as 'natural and unquestionably positive'. The *ABTS*'s counter-hegemonic strategy has used a range of communication channels to alert public opinion to the impact that the high volume of tourists is having on resource use, biodiversity and pollution, and to assert that social wellbeing and quality of life are deteriorating not improving as *Turisme de Barcelona* and other industry actors claim. Such channels include a website and a significant social media presence on platforms such as Twitter and Facebook. These are used to raise public awareness of the *ABTS*'s main concerns, advertise events and to help mobilize activist support for *ABTS*'s second 'line of action' – non-conventional forms of collective action that place pressure on public authorities to take action to curb tourism. In one of its most high profile actions in November 2016, *ABTS* activists took coordinated action to block city-tour buses in five different locations across the city to draw attention to what it described as an example of local authorities using the public purse to fund services exclusively for tourists while other bus routes popular with local people 'like the 37 or 91' were being cut (Vanguardia, 2016).

Prominent in these acts of contestation are slogans that emphasize the environmental impact of mass tourism including 'tourism kills neighborhoods', 'mass tourism is collapsing the planet' and 'the neighbour: a species threatened with extinction'. Such apocalyptic messages, that unambiguously identify tourism's unsustainability, have grabbed the wider public's attention and added to a growing awareness of tourism's negative effects. Such disquiet is reflected in opinion polls that regularly identify tourism as a major cause of public concern. In June 2017, for example, a survey carried out by Barcelona City Council revealed that tourism had 'become Barcelona's biggest problem' relegating unemployment into second place for the first time since 2008. This according to Blanchar (2016), writing in *El País*, amounted to an almost fourfold increase from 5.3% of the population in June 2015 to 19% in 2017. The *ABTS* also organizes workshops and events, which have been pivotal in the development of a series of demands that plot a course away from mass tourism, towards a model that limits the numbers of visitors to Barcelona and protects neighbourhoods and the environment. At the first Neighbourhood Forum on Tourism that took place in July 2016 under the heading, '*La ciutat es per viure-hi, no per viure'n*' ('the city is for living in, not for living off'), activists debated three main areas of concern: an economic model based on tourism, the management of ports and cruises and alternative uses for housing and public spaces (ABTS, 2016). Invitations to attend this and the second Neighbourhood Forum that took place in 2018 were extended to both domestic and international organizations engaged in struggles against tourism's excesses such as Venice's *No Grandi Navi* (No to Big Ships). These groups have now coalesced into a European Network of Cities Against Touristification (SET Network). Currently, the SET Network includes representatives from 16 major European cities including Lisbon, Venice, Malta and ten major Spanish tourism destinations such as Barcelona, Palma and Tarragona. At its launch in April 2018, the SET Network declared its intention to organize joint action to pressure governments to 'establish limits to the tourist industry' and even promote its 'degrowth' (Euractive, 2018).

The demands ensuing from these and other forums are clearly influenced by the types of wider degrowth movement concerns and policy recommendations described by Higgins-Desbiolles *et al.* (2019). They include removing public subsidies for what the ABTS (2019) describes as, 'a private activity with little positive social return such as tourism'. The ABTS (2019) is particularly critical of the funding received by *Turisme de Barcelona*, which it defines as a 'public private partnership without citizen representation'. The *ABTS* is a strong advocate of the use of taxation as a means of putting a break on tourism expansion. Thus, in addition to increasing the tourism tax that is already in place across Barcelona, the *ABTS* favours increasing taxes on the water and electricity consumed by tourists. The tourist sector, it argues, should also pay for any negative externalities that accrue to the environment and wider society as a result of its activities. It urges public authorities to use urban planning regulations to address overtourism via, for example, increasingly stringent tourism apartment and hotel licensing regimes. Finally, it proposes a debate on 'the type of city we want'. In this context, Milano (2018) argues that it is time to start talking about a city that lives 'with' and not 'from' tourism.

The *ABTS*'s 'third line of action' involves working with institutional actors to shape tourism policy in Barcelona. Institutional access has taken the form of face-to-face meetings with city council leaders and officials, attendance at public meetings organized by the city authorities to discuss tourism and involvement in deliberations leading to the co-production of important policies such as the *Plan Especial Urbanístico de Alojamientos Turísticos* (Special Urban Plan on Tourist Accommodation or PEUAT). The movement's access to institutional actors at the municipal level has been greatly facilitated by the electoral success of *Barcelona en Comú* (Barcelona in Common), a coalition of progressive social and political actors, led by the former anti-housing eviction activist and now Mayor of Barcelona, Ada Colau. Since its victory in the May 2015 municipal elections that brought Colau to power, *Barcelona en Comú* has made 'enhancing the channels of public control, citizen participation, and public–community co-production of public policies' one of its main priorities (see discussion in Blanco *et al.*, 2019).

5.6 The Political Impacts of Degrowth Contestation

In addition to impacting on public opinion and playing a role in establishing a national and international network of tourism degrowth movements, the *ABTS* has been consequential in other ways. For example, by raising the salience of tourism up the political agenda, the *ABTS* has attracted the attention of left-wing political parties that have used the issue both to demonstrate solidarity with local communities and as a means of gaining electoral support. In the Barcelona context, the party that has most identified with the issue is the left nationalist *CUP* (Popular Unity Candidacy) which has organized a number of campaigns and actions against overtourism.

As part of its electoral offer to voters in the May 2019 municipal elections, the CUP (2019) outlined several radical proposals clearly inspired by the *ABTS*'s degrowth discourse. These included reducing tourism's negative environmental impact by: limiting taxis and other private means of transport's ability to travel freely between the city and Barcelona airport and between the city and cruise ships moored in the port; reducing the carbon footprint of forms of public transport involved in tourism to a minimum; much stricter environmental controls on cruise liners than currently the case; and a hypothecated ecotax levied on both cruise liners and ferry operators to fund recovery of the marine ecosystem. In addition to the *CUP*, two other parties fighting the elections integrated *ABTS*-inspired demands into their manifestos. Both Barcelona en Comú (2019) and Esquera Republicana Catalana (2019) included fiscal measures to reduce the pressure of tourism on public spaces and services and to ensure that negative environmental externalities caused by tourism could be reduced.

The other area of influence concerns the *ABTS*'s role in the co-production of public policy. This can most clearly be seen in the context of the PEUAT and the 2020 *Plan Estratégico de Turismo para Barcelona*. In both the cases, the *ABTS* has been instrumental in shaping policies that have caused considerable disquiet among hotel chains, the *Gremi d'Hotelers*, the hoteliers' trade association, and political parties critical of *Barcelona en Comú* and its agenda of support for grassroots social actors. The hoteliers' primary concern is the measures contained in the PEAUT limiting the availability of tourist accommodation in the city (Angulo, 2017). To achieve this, the plan divides the city into four regulatory zones. The most restrictive regulation is reserved for Zone 1, which encompasses some of the neighbourhoods most densely-populated with tourist accommodation such as Ciutat Vella, the former Eixample Left, the Eixample Right, part of the district of Sant Antoni, Poble-sec, Hostafrancs, Vila de Gràcia, Vila Olympic and Poblenou. According to the plan, no new tourist accommodation can be made available in Zone 1 and if any hotels or other establishments offering overnight accommodation close, they cannot be replaced. The restrictions extend to attempts to upgrade or renovate existing structures. Thus, according to the President of the *Gremi d'Hotelers*, Jordi Clos:

> when work affects the foundations or the structure of the building, it will be necessary to reduce the number of rooms available by 20%. In zone 2, while work is allowed, it is only possible in cases where it does not lead to an increase in the overall number of rooms or in roads that are more than eight metres wide. (Angulo, 2017)

In response to the plan, the *Gremi d'Hotelers* has filed a judicial appeal against the PEAUT in the hope of getting it overturned in the Spanish courts (Cata, 2019).

5.7 Conclusion

This chapter has explored the issues of degrowth and tourism in Spain, with a focus on the Barcelona case. The main question it addresses concerns the extent of degrowth influence in shaping grassroots challenges to dominant tourism discourse in this important southern European destination, and its role in constructing alternatives to mass tourism that prioritize the interests of the environment and local communities. In order to contextualize this discussion, it set out the main contours of official tourism discourse in Spain including what the authors call a 'discourse of conviviality' that accentuates hospitality and the creation of a welcoming and safe environment for tourists. It also explored the use of the neologism, 'tourismphobia' in relation to the appearance of a polarized discourse that seeks to divide the world into 'us' and 'them' and, by so doing, legitimize the authorities' use of repressive measures to quell any forms of dissent. What is interesting, and worthy of additional exploration, is tourism activists' use of equally Manichean language as epitomized by slogans such as 'tourist you are the terrorist' that 'other' tourists as an existential threat to the city. While any detailed explanation of this phenomenon is beyond the scope of this chapter, issues meriting further exploration include the sheer depth of popular anger at the lifestyle challenge posed by hedonistic tourists staying in Airbnb type accommodation in residential parts of the city, and the influence of both left-populism and regional nationalism in activist circles in Barcelona. The other element of discourse the chapter reveals is a sustainability dimension that places leadership for sustainability firmly in the hands of the Ministry of Tourism and provides solutions to the environmental challenge mass tourism poses based on adherence to unenforceable international norms.

In its efforts to frame these and the other questions it addresses, the chapter set out a multidisciplinary framework drawing on PDA conceptualizations of ideology and the wider degrowth and tourism literature. This proved an effective vehicle for exploring both official tourism discourse and the degrowth-inspired challenge it has faced, principally in Barcelona, a city the authors consider to be the epicentre of tourism activism in Spain. Since 2000, the Catalan capital has witnessed a cycle of tourism contestation that reached its apogee in 2017. From approximately 2015, when it was set up, the degrowth-inspired *ABTS* has been at the centre of efforts to coordinate the challenge to the tourism authorities and industry in the city. In its attempt to reorder tourism in ways consistent with the principles and policy recommendations set out by the likes of Higgins-Desbiolles *et al.* (2019), the *ABTS* has adopted what it describes as a three-pronged strategy based on countering official discourse, direct action and working with institutional actors to shape tourism policy in Barcelona. Through its strategy, the *ABTS* has managed to shift public opinion about the benefits of tourism, influence important city legislation in ways that are inimical to the tourism industry's interests and influence the policy positions of important political parties such as the *CUP*, *ERC* and *Barcelona en Comú*. The chapter also demonstrated how the *ABTS* has reached out to activists in other parts of Spain and more widely to build a national and European network of anti-tourism contestation. It remains to be seen how these initiatives will evolve and whether they will prove successful in achieving a fundamental shift away from the current model based on mass tourism and growth to one based on real environmental sustainability that prioritizes the interests and needs of local communities.

References

Abend, L. (2018) Europe Made Billions from Tourists. Now It's Turning Them Away. Time, 26 July. Available at: https://time.com/5349533/europe-against-tourists/ (accessed 31 August 2019).

ABTS (2016) 1er Fòrum Veïnal sobre Turisme. Available at: https://assembleabarris.wordpress.com/forumveinalturisme/ (accessed 18 November 2019).

ABTS (2019) Resumen del Fòrum Veïnal sobre Turisme. Available at: https://assembleabarris.wordpress.com/resumen-del-2n-forum-veinal-sobre-turisme/ (accessed 10 November 2019).

Althusser, L. (1970) Idéologie et appareils idéologiques d'État (notes pour une Recherche). *La Pensée* 151, 67–125.

Angulo, S. (2017) Un centenar de hoteleros lleva el ayuntamiento a los tribunales. La Vanguardia. 11 May. Available at: https://www.pressreader.com/spain/la-vanguardia/20170511/282080571754266 (accessed 24 November 2019).

Baran, M. (2017) Trade group concerned about anti-tourism sentiment in Spain. *Travel Weekly*, 13 August. Available at: http://wwwtravelweekly.com/Travel-News/Tour-Operators/Trade-group-concerned-about-anti-tourism-sentiment-in-Spain/ (accessed 18 November 2019).

Barcelona en Comú (2019) Plans per una Barcelona Futura Programa electoral 2019–2023. Available at: https://img.beteve.cat/wp-content/uploads/2019/05/programa-electoral-ada-colau.pdf (accessed 11 November 2019).

Berger, M. and de Cleen, B. (2018) Interpellated citizens: suggested subject positions in a deliberation process on health care reimbursement. *Comunicazioni Sociali* 1, 91–103.

Blanchar, C. (2016) El 40% de los pisos turísticos de Barcelona son ilegales. *El País*, 7 September. Available at: https://elpais.com/ccaa/2016/09/06/catalunya/1473172001_414770.html?rel=mas (accessed 31 August 2019).

Blanco, I., Salazar, Y. and Bianchi, I. (2019) Urban governance and political change under a radical left government: the case of Barcelona. *Journal of Urban Affairs* 42(1), 18–38.

Campoamor, J. (2017) Turismofobia: Barcelona y otras ciudades en pie de guerra contra el turismo de masas. *BBC*, 1 June. Available at: https://www.bbc.com/mundo/noticias-40107507 (accessed 31 August 2019).

Cata, J. (2019) La justicia tumba el plan de Colau que pone coto a los hoteles en Barcelona. *El País*, 1 August. Available at: https://elpais.com/ccaa/2019/07/31/catalunya/1564567641_896396.html (accessed 18 November 2019).

Cocola-Gant, A. and Pardo, D. (2017) Resisting tourism gentrification: the experience of grassroots movements in Barcelona. *Journal of Urban Design and Planning* 13, 39–51.

Colomb, C. and Novy, J. (2016) *Protest and Resistance in the Tourist City*. Routledge, London.

Contigo Somos Democracia (2019) Posicionamientos Nacional: Turismo, Fomento, Energía. Available at: https://www.contigosomosdemocracia.es/posicionamiento-nacional/turismo-fomento-y-energia/ (accessed 15 September 2019).

CUP (2019) Programa per les eleccions municipals del 26 de maig de 2019. Available at: http://ambpasferm.barcelona/wp-content/uploads/2019/05/PROGRAMA-CUP-CAPGIREM-BCN-2019.pdf (accessed 18 November 2019).

Degen, M. (2004) Barcelona´s games: the Olympics, urban design, and global tourism. In: Sheller, M. and Urry, J. (eds) *Tourism Mobilities: Places to Play, Places in Play*. Routledge, London, pp. 131–142.

Esquera Republicana Catalana (2019) Reimpulsem Barcelona program de govern: Maragall 2019. Available at: https://img.beteve.cat/wp-content/uploads/2019/05/Programa-electoral-ERC-Barcelona-2019.pdf (accessed 18 November 2019).

Euractive (2018) Southern Europe activists join forces against mass tourism. 27 April. Available at: https://www.euractiv.com/section/economy-jobs/news/southern-europe-activists-join-forces-against-mass-tourism/ (accessed 18 November 2019).

Europapress (2018) Lanzan el programa 'Contagia tu hospitalidad' a Europa para luchar contra la 'turismofobia'. Available at: https://www.europapress.es/turismo/destino-espana/costa-blanca/noticia-lanzan-programa-contagia-hospitalidad-europa-luchar-contra-turismofobia-20180419203307.html (accessed 15 November 2019).

Freeden, M. (1996) *Ideologies and Political Theory: A Conceptual Approach*. Clarendon Press, Oxford.

Gorospe, P. (2017) Los radicales vascos dicen en San Sebastián que su lucha no es turismofobia, *El País* 17 August. Available at: https://elpais.com/politica/2017/08/17/actualidad/1502973674_417092.html (accessed 18 November 2019).

Higgins-Desbiolles, F., Carnicelli, S., Krolikowski, C., Wijesinghe, G. and Boluk, K. (2019) Degrowing tourism: rethinking tourism. *Journal of Sustainable Tourism* 27(12), 1926–1944.

Hosteltur (2017) La turismofobia en España llega a la prensa extranjera. 3 August. Available at: https://www.hosteltur.com/123380_turismofobia-espana-llega-prensa-extranjera.html (accessed 12 September 2019).

Hughes, N. (2018) 'Tourists go home': anti-tourism industry protest in Barcelona. *Social Movement Studies* 17(4), 471–477. DOI: 10.1080/14742837.2018.1468244.

Izarra, J. (2018) Retortillo da un 'ongi etorri' a los visitantes tras la campaña de la izquierda abertzale contra el turismo, *El Mundo*, 20 July. Available at: https://www.elmundo.es/pais-vasco/2018/07/20/5b51b9bd22601daa588b471d.html (accessed 18 November 2018).

La Nueva España (2017) El Gobierno ordena investigar delitos en la "turismofobia", 16 August. Available at: https://www.lne.es/asturias/2017/08/16/gobierno-ordena-investigar-delitos-turismofobia/2150151.html (accessed 12 September 2019).

Mansilla, J.A. (2018a) Vecinos en peligro de extinción, Turismo urbano, movimientos sociales y exclusión socioespacial en Barcelona. *PASOS. Revista de Turismo y Patrimonio Cultural* 16(2), 279–296.

Mansilla, J.A. (2018b) El curt estiu de la turismefòbia. Moviments socials i conflict Urb al barri del Poblenou, Barcelona. *Revista Andaluza de Antropología* 15, 74–102.

Mansilla, J.A. and Milano, C. (2018) Introducción a la ciudad de vacaciones. Apuntes sobre turismo y malestar social en Barcelona. In: Milano, C. and Mansilla, J.A. (eds) *Ciudad de vacaciones. Conflictos urbanos en espacios turísticos*. Pol.len Editorial, Barcelona, pp. 19–80y.

Mead, R. (2019) The Airbnb Invasion of Barcelona. The New Yorker, April 22. Available at: https://www.newyorker.com/magazine/2019/04/29/the-airbnb-invasion-of-barcelona (accessed 14 November 2019).

Milano, C. (2018) Saturación turística: un problema global creciente. The Conversation, 31 July. Available at: http://theconversation.com/saturacion-turistica-un-problema-global-creciente-100778 (accessed 18 November 2019).

Milano, C. and Mansilla, J.A. (2019) *Ciudad de vacaciones. Conflictos urbanos en espacios turísticos*. Pol·len Editorial, Barcelona.

Milano, C., Cheer, J. and Novelli, M. (2018) Overtourism: a growing problem', *The Conversation*, 18 July. Available at: https://theconversation.com/overtourism-a-growing-global-problem-100029 (accessed 31 August 2019).

Milano, C., Cheer, J.M. and Novelli, M. (2019) Overtourism: an evolving phenomenon. In: Milano, C., Cheer, J.M. and Novelli, M. (eds) *Overtourism: excesses, discontents and measures in travel and tourism*. CAB International, Wallingford, UK.

Oliver Sagreras, I. (2019) Turismo para un España mejor, *El País*, 18 April. Available at: https://elpais.com/elpais/2019/04/15/planeta_futuro/1555322078_863663.html (accessed 15 September 2019).

Ordíz, E. (2017) Turismofobia en España: dónde se producen las protestas y cuál es su objetivo, *20 minutos*, 13 August. Available at: https://www.20minutos.es/noticia/3111276/0/turismofobia-en-espana-claves/ (accessed 18 November 2019).

Sánchez Sánchez, E.M. (2001) El auge del turismo europeo en la España de los años sesenta. *Arbor* 170(669), 201–224. DOI: 10.3989/arbor.2001.i669.918.

van Dijk, T. (1997) What is political discourse analysis? In: Blommaert, J. and Bulcean, C. (eds) *Political linguistics*. John Benjamins, Amsterdam, pp. 11–52.

Vanguardia, L. (2016) Activistas contra el turismo masivo interrumpen por sorpresa el Bus Turístic de Barcelona, 26 November. Available at: https://www.lavanguardia.com/local/barcelona/20161126/412183821075/protesta-sorpresa-bus-turistic-barcelona.html (accessed 18 November 2019).

Ybarra, P. (2017) Los comerciantes del Casco Antiguo (Alcentro) lanzan una campaña en apoyo al turismo de Sevilla, *ABC*, 17 August. Available at: https://sevilla.abc.es/sevilla/sevi-alcentro-lanza-campana-apoyo-turismo-sevilla-201708171304_noticia.html (accessed 18 November 2019).

Zerva, K., Palou, S., Blasco, D. and Benito Donaire, J.A. (2019) Tourism-philia versus tourism-phobia: residents and destination management organization's publicly expressed tourism perceptions in Barcelona. *Tourism Geographies* 21(2), 306–329.

6 Growing Degrowth-oriented Tourism? CSR Certified Tour Operators as Change Agents

Sabine Panzer-Krause*

University of Hildesheim, Hildesheim, Germany

6.1 Introduction

Incredible growth in recent decades has led tourism to be one of the largest and most booming industries worldwide (UNWTO – United Nations World Tourism Organization, 2019). However, tourism has faced criticism since the late 1970s concerning its vast negative ecological and sociocultural impacts. The increasing number of tourist arrivals in ever-new destinations oftentimes leads to their dramatic transformation by extensive tourist infrastructure, destroying landscapes and habitats (Moscardo and Murphy, 2014, Hall, 2010). Travel-induced, high-carbon modes of transport, especially air travel, that contribute to climate change are another major problem (Hall *et al.*, 2013; Gössling and Peeters, 2015). Finally, the sociocultural consequences associated with tourism, such as all too often externally controlled tourism development, the explosion of real estate prices, labour exploitation or the inappropriate behaviour of tourists towards local customs, have evoked scepticism or even an anti-tourism position in many host communities (Adityanandana and Gerber, 2019; Milano *et al.*, 2019; Taş Gürsoy, 2019).

One effect provoked by the debate about the downsides of tourism in the past 20 years is the development of sustainability certification schemes for tourism providers, also referred to as corporate social responsibility (CSR) certification. These schemes assess tourism companies that voluntarily follow a sustainability ethos beyond political regulation and award them with a certificate if the specific sustainability requirements are fulfilled. These CSR certification schemes can be process-oriented or criteria-oriented or combine both approaches (Panzer-Krause, 2017). Besides other things, they often motivate tourism companies to comply with elements that can be categorized as degrowth-inspired aspects of tourism. Andriotis (2018) describes this

*sabine.panzer-krause@uni-hildesheim.de

DOI: 10.1079/9781789245073.0006

sufficiency-driven form of tourism as one that enables travellers to engage with a destination's natural and cultural heritage, while empowering local communities and reducing CO_2 emissions by using low-carbon means of transport.

With regard to a degrowth-oriented transition of the tourism industry, for which more and more scholars plead (Hall, 2010; Andriotis, 2018; Fletcher *et al.*, 2019; Higgins-Desbiolles *et al.*, 2019), CSR-certified tourism companies may function as change agents. In particular, tour operators could play an important role since they act as intermediaries and select different individual tourism offers along the tourism supply chain and bundle them into packages in order to sell them to their customers. Thus, they have a substantial influence on the tourism sector. This is underpinned by the fact that in Germany, for example, 43% of all vacation trips are booked as package or modular tours organized by tour operators (FUR – Forschungsgemeinschaft Urlaub und Reisen, 2019).This study seeks to add to the literature concerning the transformation of the tourism sector towards genuinely sustainable modes. Using an evolutionary approach as a theoretical lens, it analyses whether German tour operators who have gained the CSR certification 'TourCert' and have committed themselves to harmonizing ecological, sociocultural and economic goals concerning their tour planning operations have the potential to upscale the downscaling idea of degrowth in tourism to embrace more of the tourism market and thus to further a reformist pathway to structural change.

The chapter is structured as follows. Section 2 forms the chapter's theoretical basis, Section 3 outlines the methodology of the study, Section 4 presents and discusses the results of the analysis and Section 5 offers a conclusion.

6.2 Degrowth in Tourism?

6.2.1 Degrowth as an alternative for a booming sector?

The tourism industry has been massively growing worldwide for the last seven decades with international tourist arrivals increasing from approximately 25.2 million in 1950 to about 1.4 billion in 2018. Tourism receipts amounted to 1.4 trillion US dollars in 2018 (UNWTO, 2019). As a consequence, many destinations like the coastal regions of the Mediterranean have experienced a stampede of tourists and quickly developed into mass tourism destinations with rapid and mostly unplanned urbanization accompanied by dramatically negative environmental and sociocultural impacts since the 1950s. Tourism in such mass tourism destinations is large scale and the tourists travelling to these regions are usually considered 'sun, sand and sea' tourists who are physically rather passive and do not show much interest in the region's landscape and culture (Bramwell, 2004; Blázquez-Salom *et al.*, 2019).

In the late 1970s, criticism of this type of tourism became louder and small-scale tourism as an alternative to mass tourism was advocated. In contrast to mass tourism, small-scale tourism is, ideally, developed by local people and thus locally controlled. The aim is to prevent drastic changes and thus severe damage to natural and cultural heritage, and to ensure that benefits remain within the local context. Small-scale tourism is often located in rural and economically underdeveloped regions, and in many rural areas this form of tourism is used as a conservation tool to protect the natural and cultural countryside (Powell and Ham, 2008; Cortes-Vazquez, 2017). Hence, small-scale tourism is closely linked to sustainable development and is seen by many scholars and practitioners as the 'antithesis of mass tourism' (Lane and Kastenholz, 2015, p. 1139). Nevertheless, the countryside has experienced increasing popularity, too, and following capitalist logic, there has also been an immense growth of small-scale tourism (Becken, 2005; Paunović and Jovanović, 2017; Panzer-Krause, 2020).

Meanwhile, ever-new destinations emerge on the international tourism horizon oriented somewhere between the two extremes of large-scale and small-scale tourism, but generally growth-driven. Despite at times high leakage with respect to local value added, for many destinations tourism is an important, often even the most important, economic pillar. Also, UNWTO (2019) regards tourism as a favourable option to promote sustainable development, especially for poorer communities in the Global South. Growth-oriented sustainable tourism is considered a catalyst for development and positive change. Concerns about the ecological and sociocultural capacity of destinations are dismissed by highlighting that the quantity of tourists is not the problem; it is rather the case that new strategies for adequately managing the increasing tourist numbers are required (Rifai, 2017).

However, there has been increased questioning of such green growth strategies that claim to be able to combine sustainable development and economic growth. Following scholars who declare that continuous growth will lead to collapse due to the planet's finite ecological and economic capacity (Jackson, 2009; Latouche, 2009; Kallis, 2018), Andriotis (2018), among others, argues that a degrowth-oriented approach needs to be applied in the tourism sector in order to rightsize or downscale tourism activity. Although he notes that there is no coherent understanding of the key aspects of degrowth in a tourism context and that the concept lacks clarity and precision, he identifies several characteristics that he links to degrowth-inspired travelling from a consumer perspective. Accordingly, degrowth-inspired travelling is characterized by its low-carbon means of transport, as for example used by backpackers who do not attach value to quickly getting from one point of interest to the next. Furthermore, degrowth-inspired tourism 'embraces philosophies of life reflecting non-materialistic sources of satisfaction and a minimally "sufficient" material standard of travelling' (Andriotis, 2018, p. 192). In this sense degrowth-oriented tourism concentrates on non-mainstream travellers who seek to see unknown places that are not crowded with tourists and offer an unspoilt environment. The tourists who visit such hidden gems request only simple facilities such as hostels, camping sites or a public transport system. Generally, they reject being objects of commercialization (Andriotis, 2018). Degrowth-oriented tourism instead aims to bring more benefits to local communities by following the idea of empowering civil society and aspects of commoning (Büscher and Fletcher, 2017; Adityanandana and Gerber, 2019; Fletcher *et al.*, 2019; Navarro-Jurado *et al.*, 2019). Here, Chassagne and Everingham (2019) as well as Renkert (2019) see parallels with the concept of *buen vivir* which originates from the indigenous communities of Latin America and prioritizes the growth of social and environmental wellbeing instead of economic growth. However, to a large extent it remains unclear what degrowth in tourism and the pathway to it might look like.

With regard to these principles, it can be ascertained that the degrowth tourism approach assembles key ideas of concepts such as rural tourism (Panzer-Krause, 2019), slow tourism (Heitmann *et al.*, 2011), staycation (James *et al.*, 2017), responsible tourism (Leslie, 2012), eco-tourism (Fennell, 2015), heritage tourism (Park, 2013) and religious tourism (Griffin and Raj, 2017), and also adds the note of a general need for less production and consumption.

Yet, as Fletcher *et al.* (2019) and others stress, degrowth-oriented tourism is not to be mistaken for an anti-tourism position. Instead, it pleads for tourism to be differently scaled and organized. In this context, Gascón (2019) condemns the frivolity of many tourists and stakeholders who claim that tourism is a human right.

Beyond this, Hall's (2009, 2010) calls for a steady-state understanding of tourism and urges that tourism needs to be integrated into a circular economy, which Daly (1991 [1977]) defines as a non-growth society with a stable throughput of resources. Here, degrowth as a form of rightsizing the tourism industry is necessary to move towards this equilibrium, in which, Hall's (2009, 2010) argues, both the degrowth-oriented sufficiency approach and the green growth-oriented efficiency approach as sustainability strategies are required.

With regard to realizing degrowth in a booming sector, numerous scholars demand a radical political–economic reorganization which embeds tourism into a post-capitalist system (Fletcher *et al.*, 2019; Higgins-Desbiolles *et al.*, 2019). This transformation is characterized as a civil society-based triple movement 'with pushes for systemic structural change at global and societal levels combined with more localized and individual practices contributing to downsizing on the ground' (Fletcher *et al.*, 2019, p. 1754). Blázquez-Salom *et al.* (2019) support this radical view. In a Spanish study they observe that pursuing degrowth within the capitalist system by means of market-led territorial and tourism planning instruments only leads to an unfavourable exclusive and elitist use of space in tourism destinations rather than to socioenvironmental justice.

However, in a world where a sudden disappearance of the growth ethos seems to be unrealistic in the foreseeable future and where, especially in western societies 'policy makers will avoid taking unpopular decisions' (Font, 2017, p. 210), a transition of the booming tourism industry to a more degrowth-oriented sector seems to require a less radical evolutionary approach to change. Against this background, a successful attempt to degrow tourism largely depends on its capacity to upscale degrowth-inspired travel from a niche to a wider context.

While the next section lays the theoretical foundations for an evolutionary approach to foster tourism degrowth without radical change, the empirical study explores the practical capacities for growing degrowth-inspired travel in the tourism sector. In contrast to Andriotis (2018), here the focus is on the producer side.

6.2.2 Strategies for a degrowth transition and tour operators as change agents

Scholars advocating degrowth-inspired travelling associate this form of tourism with characteristics such as non-mainstream, small-scale and less consumptive (Andriotis, 2018; Fletcher *et al.*, 2019; Higgins-Desbiolles *et al.*, 2019). However, Weaver (2012) doubts that small-scale and non-mainstream tourism is applicable for the whole sector and strives for a pragmatic approach with regard to developing strategies to establish sustainable mass tourism destinations. Here, management approaches developed and tested in isolated rural tourism hot-spots that are surrounded by an otherwise small-scale tourism sector can offer test fields for acquiring new practices (Panzer-Krause, 2020).

This being said, a great variety of sustainability strategies can be identified. While they are all mostly contextualized between the two extremes of the efficiency-driven green growth approach (von Weizäcker *et al.*, 2009; UNWTO, 2019) and the sufficiency-oriented degrowth approach (Jackson, 2009; Latouche, 2009; Kallis, 2018), many scholars agree that sustainability has to be regarded as a process (Faber *et al.*, 2005). Consequently, the transformation of the tourism industry towards a genuinely sustainable sector with degrowth-oriented travelling as a key element in order to rightsize the industry has to be guided by an evolutionary approach that involves exploring and analysing reformist pathways that lead to sustainable change.

Degrowth scholars bemoan that the common understanding of the term 'sustainability' is closely linked to green growth (Fletcher, 2011; Higgins-Desbiolles, 2018; Fletcher *et al.*, 2019; Higgins-Desbiolles *et al.*, 2019). However, based on the general degrowth discourse, they argue that radical change of the tourism sector and the entire capitalist conception as such is necessary in order to achieve a paradigmatic shift towards degrowth-oriented tourism. According to this view, degrowth within the capitalist system marks an 'impossibility theorem' (Foster, 2011) which can only be overcome by a type of post-capitalism that rejects the general growth ethos. Yet, currently it seems unrealistic to think of finding majorities for radical change to the capitalist system. On the one hand, this would mean western societies dispensing with economic hegemony. On the other hand, many societies in the Global South give priority to their economic development (Foster, 2011). Hence, under the current circumstances and despite all the challenges, strategies for non-radical, reformist change within the current capitalist system need to be explored and theoretically substantiated.

Ma and Hassink (2013) propose applying the evolutionary approach in the tourism sector as developed in evolutionary economic geography, a new paradigm in contemporary economic geography (Boschma and Martin, 2010). This theoretical lens provides an opening to investigate the rise and fall of tourism destinations. In contrast to Butler's (1980) deterministic tourism area life cycle model, the evolutionary approach takes an actor-oriented perspective and thus enables theoretical explanations for diverging pathways of tourism destinations. Panzer-Krause (2017, 2019) adopts this evolutionary approach to theoretically unravel the pathways and coevolutionary processes concerning sustainability transitions in the tourism industry. Brouder (2017) also recognizes potential for explaining sustainability transitions in the tourism sector by means of an evolutionary lens.

Voluntary CSR certification that goes beyond political regulation has become a strategy for tourism companies to get engaged in sustainable change and communicate their sustainability commitment. During the last two decades, several hundred certification schemes have been established worldwide (Balš and Strasdas, 2015). Panzer-Krause (2017, 2019) suggests that such certification schemes applied in specific destinations might be able to initiate new trajectories for more sustainable modes of tourism development. In two empirical studies she investigated networks of tourism businesses that voluntarily and regularly undergo a CSR certification process in order to comply with a general sustainability ethos and follow certain ecological, sociocultural and economic aspects in their business activities (Panzer-Krause, 2017, 2019). The studies were realized on the island of Rügen, Germany, and the Burren, Ireland, which represent two vulnerable destinations that are under pressure from drastic tourism development and that show signs of overtourism due to ecological and sociocultural capacity limits. Although the two certification schemes can be regarded as utterly low-threshold they established quite different certification processes and internal structures. Nevertheless, in both local tourism networks a spectrum of different sustainability orientations among the actors was identified that ranged from strict green growth approaches to rigid degrowth approaches. Panzer-Krause (2019) argues that degrowth strategists need to be key actors in innovation networks of the tourism sector to realize a clear shift away from business as usual. However, she also highlights that the 'sustainability light' approach followed in the two destinations raises concerns about the pace of sustainability transitions and possible time limits that need to be acknowledged in order to avoid severe socioenvironmental disturbance.

So far, the focus of academic studies regarding the development of more sustainable modes of tourism has been mainly destination-centred. The number of destinations where networks of tourism businesses have been established that follow a sustainability ethos and promote sustainable tourism within the local context has risen significantly during the last decade. Nonetheless, the impact of regional efforts to transform tourism within specific destinations is spatially limited. Additionally, most tourists travelling to these areas are not aware of their existence (Balš and Strasdas, 2015).

For this reason, this study aims to consider the whole value chain of tourism product creation. CSR certified tour operators that bundle tourism services along the tourism supply chain might be more influential with regard to a change towards more degrowth-oriented tourism. As noted above, in Germany, nearly half of all vacation trips are organized by tour operators (FUR, 2019). Thus, with regard to a reconfiguration of tourism, tour operators may act as change agents.

6.3 Methodology

6.3.1 The TourCert certification scheme

For this study, TourCert, a German-based and internationally recognized non-profit certification system for sustainable tourism, was used to investigate whether CSR certified tour operators act as change agents for a transition towards degrowth-inspired tourism. TourCert was founded in 2009. In 2019 the certification scheme had reached 330 certified tourism companies and destinations all over the world, especially in Germany and Latin America (TourCert, 2019a). TourCert considers the whole tourism value chain, hence tour operators that bundle tourism offers into packages are, with 195 certified operators in 2019, the key target group. It is a primarily process-oriented certification scheme and operates in conformity with international standards, developed for example by the Global Sustainable Tourism Council (GSTC) (TourCert, 2019b).

According to its process orientation, which generally hides the risk of greenwashing (Panzer-Krause, 2017), there are only a few requirements for the initial certification, which mainly consist of transparent sustainability reporting including ecological, sociocultural and economic indicators. However, annual improvement targets and action plans have to be set by every certified partner. After two and for each further period after three years, the businesses have to be recertified (TourCert, 2019b).

In a study carried out by Die Verbraucher Initiative e.V. and ZENAT (2016), TourCert certification for tour operators was ranked the best German sustainability certification scheme in the tourism industry among the 36 certification systems analysed.

6.3.2 Data collection and analysis

The empirical study was realized by documentary and desktop research. Fifty-nine German tour operators that held full TourCert certification in 2019 were identified (ID 1-ID 59). The study drew on the tour operators' detailed sustainability reports or alternatively the summarized audit reports, which mainly comprise key sustainability indicators. Certified tour operators with at least five full-time equivalent employees are supposed to publish their reports on their websites in order to ensure transparency (TourCert, 2015).

However, only the reports of 26 of the 59 tour operators were accessible online. The other companies were contacted and invited to transmit their reports. Thereupon, another ten reports of nine businesses were received. Twenty-one tour operators did not react, and five businesses explained that due to the small size of their business, either a formal CSR certification had not been conducted or they were not obliged to hand over their report to a third party. Altogether, 37 sustainability and 14 audit reports of a total of 35 TourCert-certified tour operators could be included in the analysis.

The reports had very different elaboration dates. While the most recent reports of 20 businesses were up to date, the most current reports of 15 tour operators had been written before 2017, even though recertification is usually due after two to three years and thus reports should not be older than that. Furthermore, for 26 of the 35 businesses under investigation only the most current report was accessible. Hence, the reports included in the study drew on data generated between 2010 and 2018. The poor data situation complicated both a process-oriented and a comparative approach to the analysis.

To probe the TourCert certification process as well as the tourism companies in order to identify degrowth-inspired elements of tour organization and evaluate their potential for upscaling, the reports were used for both a quantitative and a qualitative analysis. As far as possible the tour operators' key sustainability indicators were extracted and compared. These included indicators such as the CO_2 emissions per guest per night, the value-added share that remains within the toured destinations, a sustainability index concerning the product design, a client satisfaction index, a sustainability index concerning partner agencies, a sustainability index concerning accommodation and a sustainability index concerning tour guides. Additionally, the descriptive sustainability reports, where available, were investigated qualitatively following Mayring (2015), by using categories such as 'ambitions for business growth', 'choice of destinations, mode of travel and CO_2 compensation', 'cooperation with suppliers' or 'community involvement'. The aim of this qualitative content analysis was to summarize, explicate and structure the sustainability reporting material.

The study was complemented by means of an analysis of relevant documents of industry-specific German associations and research organizations such as Forum anders reisen (2019) and Forschungsgemeinschaft Urlaub und Reisen.

6.4 CSR Certified Tour Operators Pushing Towards Degrowth-oriented Tourism

6.4.1 Characteristics of the German TourCert certified tour operators

Fifty of the 59 German tour operators that are certified with TourCert are simultaneously members of Forum anders reisen, the German association of sustainable tour operators. Guidelines and criteria for sustainable travel

Table 6.1. Full-time equivalent employees (*N*=33) by TourCert operator.

Number of full-time equivalent employees	1–<5	5–< 10	10–<20	20–<50	50–<100	100+
Number of TourCert tour operators	16	5	4	4	2	2

organization are worked out within the association. The TourCert certification scheme also emanated from Forum anders reisen, and members are encouraged to gain certification (Forum anders reisen, 2019).

The TourCert certified tour operators under investigation in this study are mainly very small or even micro-sized businesses. Sixteen were identified as businesses with less than five full-time equivalent employees, another five had between five and under ten full-time equivalent employees. Only two employed between 50 and 99 people, and another two had more than 100 employees (Table 6.1). Six operators served less than 1000 travellers in their annual reporting periods. Another five businesses served between 1000 and less than 10,000 tourists. Only two businesses planned the travel of more than 50,000 tourists (Table 6.2).

With respect to the destinations selected for travelling, 27 of the 35 CSR certified tour operators focus their business activities mostly on destinations outside mainland Europe, many of them being rather exotic and remote destinations such as Trinidad and Tobago, Madagascar, Mongolia or Oman, at which 14 of the tour operators specialized in only one or a few very specific destinations.

While, comparable to this study, in 2018 about 94% of the approximately 1500 German tour operators had less than 50 employees (Kirstges, 2018), the destinations preferred by German tourists differed considerably with Spain being the top international destination and only 8% of all trips being long-distance trips (FUR, 2019). In addition, the nearly 50 million trips sold by German tour operators in 2018 (Kirstges, 2018) also suggests that TourCert certified tour operators represent only a small part of the German tour operating market.

6.4.2 Transparency of the certification scheme

Die Verbraucher Initiative e.V. and ZENAT (2016) attested a very high transparency rate of 92% for the TourCert certification scheme for tour operators,

Table 6.2. Travellers served annually (*N*=24) by TourCert operator.

Travellers served annually	<1000	1,000–<10,000	10,000–<20,000	20,000–<50,000	50,000+
Number of TourCert tour operators	10	8	1	2	3

even though there remains little clarity with respect to how this value was calculated. However, the present study raises concern regarding the scheme's transparency for the following reasons.

First, for consumers, transparency is limited with regard to general access to tangible CSR-related information as the certification scheme has developed different options for certified businesses to leave their customers in the dark. Thus, micro-sized enterprises with less than five full-time equivalent employees only have to complete a simplified certifying and recertification process and are not obliged to publish any reports or indicators (TourCert, 2015). On the one hand, a slim certification process for these businesses seems sensible with regard to the explanatory power of certain indicators for them and the resources absorbed with respect to data collection. On the other hand, these operators do not usually reveal their business size to the public which makes it an unclear policy from the perspective of the customer. Furthermore, businesses with at least five full-time equivalent employees, which are obligated to publish their certification results, can waive up to three indicators with which they are not content. Finally, it is not clear why some of the reports that are published are older than three years although recertification is due after this time.

Second, further incoherence pertains with regard to the ten TourCert CSR indicators that are supposed to illustrate business performance. These include, for example, the value-added share that remains within the toured destinations or the carbon emissions per guest per day, and in addition also indices comprising several indicators of a certain field, e.g. product design. However, neither TourCert nor the businesses' reports themselves outline what data flows into the calculation of these indices or to what extent. Also, when measuring the key indicators, tour operators may consider only a certain segment of their business activities in which a larger share of the turnover is generated (TourCert, 2015, 2018). Here, the data chosen to be included in the analysis may vary from assessment to assessment.

Third, a significant lack of transparency is evoked by the heart of the TourCert certification: its process-oriented character. On the basis of its key indicators, each certified business develops an action plan with targets for the time until recertification is due in order to continuously improve its CSR performance (TourCert, 2015). Yet, it remains unclear what a good indicator value is and what the minimal requirements are in order to gain certification. Moreover, an improvement in CSR performance does not seem to be necessary for recertification. Two of the tour operators, for which at least two data sets for different reporting periods were available, were even recertified despite a massive deterioration in their CSR performance regarding several indicators. Additionally, no information is given about the consequences if measures of the action plan are not realized within the specified timeframe.

Especially for micro-sized businesses, the time and effort needed for the TourCert certification process are considerable in relation to the resources they can provide. Yet, this group of tour operators forms the main TourCert certified faction, hence special arrangements for them diminishes the relevance of the certification scheme and thus also encourages certified tour operators to do 'business as usual' (Ferguson, 2014) rather than push for a shift towards degrowth in tourism.

6.4.3 The paradox of growing degrowth

The sustainability or audit reports of many of the 35 tour operators under investigation explicitly reveal their intention for business growth:

> 'The overall very positive past years' development of sales continues in 2015. We expect sales to grow more than 20%' (ID 1 2015: 16).
>
> '[We] continue to consistently follow our business growth path' (ID 35 2017: 5).
>
> 'The business has continuously grown since its inception … In the current reporting year it is also worth mentioning that new product lines in the field of group tours were established with the aim of expanding our offerings and especially of reaching new target groups' (ID 39 2016: 7).

This growth ethos contradicts the idea of degrowth in tourism. However, in the sense of Schumpeterian creative destruction (Schumpeter, 2008 [1942]), it may still lead to a restructuring of the tourism market if innovative elements of degrowth-inspired travel are realized in the certified businesses' tour operating activities and if these have the potential to be upscaled to wider parts of the market and thus force out conventional tourism offers. In the following, this will be evaluated with regard to the modes of transport on organized tours and the holiday design at the destinations travelled to.

Forum anders reisen (2019) set up guidelines for the length of stay at a destination depending on the distance that needs to be covered by air travel. These guidelines were incorporated by TourCert. Accordingly, for one-way flight distances of more than 3800 km, tour operators should adhere to a length of stay of at least 14 days for their planned tours. With regard to flight distances between 800 and 3800 km, holidays should last one week as a minimum. Air travel to destinations less than 800 km away should be avoided. The analysis shows that TourCert certified tour operators normally comply with these guidelines. Nevertheless, their holidays, even in far-away and remote destinations, usually do not last longer than three weeks. While Forum anders reisen (2019: 4) refers to a 'justifiable relation' between distance and length of stay, the criteria for this remain unclear.

In accordance with the policy of Forum anders reisen (2019) adopted by TourCert, the certified businesses' mean air kilometres per tourist for a three-week holiday can be as high as 23,137 (ID 4). As this obviously contributes neither to sustainable change nor a positive image for a CSR certification scheme, the certified tour operators are additionally encouraged to compensate CO_2 emissions generated by their clients' air travel. As suggested by Forum anders reisen (2019) and likewise adopted by TourCert, most of the tour operators inform their clients about the possibility of CO_2 compensation and leave the decision and the costs up to them:

> 'Each client is informed about the carbon balance of a tour during the booking process and has, thanks to our cooperation with atmosfair, the possibility to compensate the trip's carbon production financially' (ID 4 2017: 20).

> '[We cooperate] with the climate protection initiative atmosfair and recommend our customers donate to atmosfair for the CO_2 emissions they cause' (ID 14 2015: 5).

> 'We design our tours to be as climate-friendly as possible and offer our customers the possibility for compensation by means of a donation to atmosfair' (ID 38 2016: 9).

Only two tour operators fully and nine partly include the costs of CO_2 compensation in their travel packages. For this policy, one tour operator was even awarded the atmosfair Bronze Award in 2010 and 2011 (ID 31).

Increasing awareness of CO_2 compensation by the tour operators' customers has allowed all nine tour operators under review, for which at least two sets of data were accessible, to reduce their CO_2 footprint per guest per night over the last few years. However, this can only be considered a cosmetic recalculation, not a genuine reduction of CO_2 emissions in the sense of degrowth, and has rather to be referred to as a strategy of greenwashing. In contrast, a degrowth-inspired reduction of CO_2 emissions can be realized in two ways. First, when resorting to air travel as a mode of transport, tour operators can choose destinations that are closer to the customers' country of origin in order to diminish air kilometres. However, for many tour operators a restructuring of their often very specific destination profiles is not up for debate:

> '... the analysed tours are exclusively long-haul trips whereby the long distances have to be covered by aeroplane' (ID 13 2018: 14).

> 'In view of the destinations offered, clients mainly travel by aeroplane' (ID 35 2017: 17).

> 'As a provider of long-haul tours, unfortunately, air travel is indispensable for our business' (ID 39 2016: 20).

Second, tour operators can opt for other modes of travel which cause less emissions such as buses, trains, sailing boats or bicycles. With respect to the average distances travelled, for many overseas destinations this is not a practicable alternative:

> 'As our travel areas are located far away, there is mostly no realistic alternative to air travel' (ID 11 2015: 5).

> 'Due to the long distance to our destinations, there is virtually no alternative to travelling by aeroplane' (ID 27 2019: 2).

Yet, there are 2 of the 35 TourCert certified businesses under investigation that organize home stays in other countries with the purpose of learning a new language and cultural exchange. In contrast to most other journeys organized by the tour operators, these are indeed more degrowth-oriented if they include longer stays. In sum, however, although long-distance travel is one of the characteristics of the TourCert certified tour operators, upscaling it is not in line with a degrowth-oriented restructuring of the tourism market. Here, the CSR certified operators under investigation are not in a position to act as change agents.

With regard to the holiday design at the destinations, TourCert certified tour operators realize numerous aspects of specific forms of small-scale tourism such as slow tourism, ecotourism or rural tourism. The operators, although to different extents, cooperate with local partner agencies, accommodation and tour guides and aim to avoid external providers at the same time. Five operators could be identified that even partly or fully dispense with hotels and accommodate their clients in host families instead, although three of them only do this as a tourist highlight for one or two nights. In the sense of commoning tourism (Fletcher *et al.*, 2019), this measure ensures an authentic holiday experience for the tourists as well as generating benefits for the host communities. Altogether, significantly higher amounts of value added remain within the destinations and capital drain to external, international companies is reduced. In addition, five TourCert certified tour operators put emphasis on selecting CSR certified accommodation providers. Another 16 that work with a specific supplier code of conduct could be identified. Furthermore, within the destinations, the tour operators try to use environmentally friendly modes of transport which also benefit local communities, such as public buses and trains, taxis, bicycles and cycle rickshaws or boats. Hence, in this respect, they mostly act as change agents and spread their vision of sustainability to more tourism actors along the tourism supply chain. Nevertheless, 12 of the 35 tour operators under investigation said they need to fall back on domestic flights in order to get from one holiday region to the next.

The holiday experiences of the tour operators under investigation mostly include active elements such as walking, cycling or getting in touch with local communities, but simultaneously renounce activities like helio-skiing or quad-driving that are environmentally harmful (Forum anders reisen, 2019). As tour guides are encouraged to pass on sustainability information, the holiday product is additionally enriched with varying amounts of information about natural and cultural heritage. While these are elements of degrowth-inspired tourism, it is not desirable that CSR certified tour operators who travel to remote and so far unspoilt places function as change agents in terms of transferring them to the mass market, since managing large quantities of tourists in vulnerable natural and cultural landscapes is beset with many difficulties. An increasing number of rural tourism hot-spots around the world already struggle with these challenges of overuse (Panzer-Krause, 2020). Apart from that, the CSR certified tour operators consciously distance themselves from mass tourism and consider their tourism activities as niche products:

> 'Class instead of mass' (ID 38 2016: 9).
>
> ' … a specialist for individual travelling far away from mass tourism' (ID 47 2015: 6).

Finally, community involvement at the destinations forms an important part of many TourCert certified tour operators' activities. While 24 of the tour operators financially support a wide range of community and environmental projects and thus demonstrate their commitment to promoting socioenvironmental justice at the destinations indirectly, four even enable their clients to get involved in certain projects themselves. The former is applicable to the mass tourism market, and CSR certified tourism businesses can be change agents to diffuse this practice. With regard to the latter, however, the success of this form of volunteer tourism is critically debated (Forum anders reisen, 2019). Additionally, upscaling it to the wider tourism market is neither favourable nor feasible.

The analysis of the tour organizing activities of TourCert certified tour operators exposes the paradox of growing degrowth-oriented tourism. Some elements of degrowth-inspired tour operating can be identified that have the potential to be upscaled to wider parts of the market and entail positive effects of rightsizing, such as close cooperation with local partners and the indirect support of environmental and community projects. However, there are a number of practices that would lead to contrary and thus vastly negative effects, if widely diffused. These include long-distance air travel and visiting vulnerable, so far unspoilt places, as is popular within this sector, and active involvement in vulnerable communities and environments. Even with regard to the niche market CSR certified tour operating represents at the moment, this needs to be critically reflected upon. At the same time, it is doubtful whether the general growth ethos the CSR certified businesses have internalized will lead to creative destruction as described by Schumpeter (2008 [1942]) and thus to a restructuring of the tourism market.

Hence, with a view to the paradox of growing degrowth-oriented tourism, either the demand to restructure the tourism industry of entire societies or the approach of degrowth in tourism itself need reconsideration. On the one hand, reorganizing tourism in accordance with a non-mainstream concept entails selective and thus in some way elitist access to travel. On the other hand, upscaling degrowth with regard to the characteristics associated with it currently will not lead to rightsizing the tourism industry.

6.5 Conclusion

Using an evolutionary approach (Ma and Hassink, 2013) and thus acknowledging that sustainability needs to be regarded as a process (Faber *et al.*, 2005), this study has evaluated the potential of CSR certified tour operators in Germany to act as change agents and grow degrowth-inspired tourism. Hence, the study followed Hall's (2009, 2010) argument of a circular economy in the tourism sector which requires initial downsizing by means of degrowth-inspired tourism (Andriotis, 2018; Fletcher *et al.*, 2019; Higgins-Desbiolles *et al.*, 2019).

Referring to German tour operators that have gained the CSR TourCert certification, the study analysed the operators' sustainability or audit reports, as far as these were accessible, by means of a qualitative and quantitative analysis. The analysis revealed that the TourCert certified tour operators follow various measures, although to different extents, that are in line with degrowth-inspired travelling as described by Andriotis (2018). This especially applies to elements concerning the sociocultural dimension of tourism, including close cooperation with local partners. By continuously increasing their engagement in these fields and communicating them to the whole sector, CSR certified operators can indeed act as change agents, pushing towards a more sustainable tourism industry and even some aspects of commoning. However, especially concerning the choice of destinations and the selected mediums of transport in order to reach them, TourCert operators, in general, do not set a good example. They tend to choose rather remote destinations which require long-distance travel, in most cases realized by air. While they are typically not willing to reconsider their tour destinations, they do opt for CO_2 compensation, but mainly only by informing their clients about this option. Attempts to upscale these practices of choosing destinations and mediums of transport for a greater target group would not be a suitable pathway in terms of rightsizing tourism. On the contrary, this would quickly lead to environmental collapse. This paradox of growing degrowth implies that a clear differentiation between the mass market and a small-sized alternative travel sector is necessary. It is a contradiction to attempt to upscale small-sized tourism for sustainability purposes. However, it remains to be seen whether the strict travel restrictions temporarily put in place by nations all over the world in the course of the COVID-19 pandemic in 2020 lead to a radical change in the tourism industry. This may especially include a profound restructuring of the international airline sector on the one hand and a stronger desire of tourists to choose nearby holiday destinations on the other.

Furthermore, this study shows that the German TourCert certified tour operators only represent a very small niche of the whole tourism market. The majority of the businesses under investigation serve less than 10,000 clients annually. Compared to the nearly 50 million trips sold by German tour operators in 2018 (Kirstges, 2018) this is extremely small. Thus, their potential to restructure the tourism industry in Germany towards degrowth is very limited. Additionally, the TourCert certification scheme lacks a great deal of transparency since there are special arrangements for micro-sized tour operators, the main faction of certified businesses. This rather encourages the tour operators to do 'business as usual' (Ferguson, 2014) without feeling guilty about it. Here, 'best practice' is constructed by a niche sector that generally claims to be more sustainable than the rest of the industry (Valdivielso and Moranta, 2019).

In sum, it has to be acknowledged that current CSR certification schemes do not seem to genuinely foster the restructuring of the tourism market within the capitalist system, but can only marginally advocate and diffuse certain elements of degrowth-oriented tourism. On the one hand, CSR certification schemes rather lack the influence necessary for a paradigm shift. On the other hand, at present the approach of degrowth-oriented tourism seems unsuitable for mainstream application.

Thus, the question remains open as to how the tourism industry as such can be transformed towards a steady state economy (Hall, 2009, 2010) within the capitalist system. If the COVID-19 pandemic does not lead to a profound restructuring of the international travel sector, a radical renunciation of tourism seems unrealistic in the foreseeable future and thus, non-radical change continues to be inevitable.

References

Adityanandana, M. and Gerber, J.-F. (2019) Post-growth in the tropics? Contestations over *Tri Hita Karana* and a tourism megaproject in Bali. *Journal of Sustainable Tourism* 27(12), 1839–1856. DOI: 10.1080/09669582.2019.1666857.

Andriotis, K. (2018) *Degrowth in tourism. Conceptual, theoretical and philosophical issues*. CABI Publishing, Wallingford/UK, Boston, Massachusetts/USA.

Balš, M. and Strasdas, W. (2015) Nachhaltiges Tourismusmanagement. In: Rein, H. and Strasdas, W. (eds) *Nachhaltiger Tourismus. Einführung*. utb, Stuttgart/Germany, pp. 231–272.

Becken, S. (2005) The role of tourist icons for sustainable tourism. *Journal of Vacation Marketing* 11(1), 21–30. DOI: 10.1177/1356766705050840.

Blázquez-Salom, M., Blanco-Romero, A., Vera-Rebollo, F. and Ivars-Baidal, J. (2019) Territorial tourism planning in Spain: from boosterism to tourism degrowth? *Journal of Sustainable Tourism* 27(12), 1764–1785. DOI: 10.1080/09669582.2019.1675073.

Boschma, R. and Martin, R. (eds) (2010) *The Handbook of Evolutionary Economic Geography*. Edward Elgar Publishing Limited, Cheltenham, UK.

Bramwell, B. (2004) Mass tourism, diversification and sustainability in Southern Europe's coastal regions. In: Bramwell, B. (ed.) *Coastal mass tourism: Diversification and sustainable development in Southern Europe*. Channel View Publications, Clevedon/UK, et al., pp. 1–21.

Brouder, P. (2017) Evolutionary economic geography: reflections from a sustainable tourism perspective. *Tourism Geographies* 19(3), 438–447. DOI: 10.1080/14616688.2016.1274774.

Butler, R.W. (1980) The concept of a tourist area cycle of evolution: implications for management of resources. *The Canadian Geographer/Le Géographe canadien* 24(1), 5–12. DOI: 10.1111/j.1541-0064.1980.tb00970.x.

Büscher, B. and Fletcher, R. (2017) Destructive creation: capital accumulation and the structural violence of tourism. *Journal of Sustainable Tourism* 25(5), 651–667. DOI: 10.1080/09669582.2016.1159214.

Chassagne, N. and Everingham, P. (2019) Buen Vivir: Degrowing extractivism and growing wellbeing through tourism. *Journal of Sustainable Tourism* 27(12), 1909–1925. DOI: 10.1080/09669582.2019.1660668.

Cortes-Vazquez, J.A. (2017) The end of the idyll? Post-crisis conservation and amenity migration in natural protected areas. *Journal of Rural Studies* 51, 115–124. DOI: 10.1016/j.jrurstud.2017.02.005.

Daly, H.E. (1991 [1977]) *Steady-state economics*. Island Press, Washington, D.C./USA.

Die Verbraucher Initiative e.V. and ZENAT (2016) Anforderungen an Unternehmenszertifizierungen für nachhaltigen Tourismus in Deutschland. Available at: http://www.zenat-tourismus.de/images/pdf/Ergebnisbericht_Nachhaltiger-Tourismus.pdf (accessed 27 December 2019).

Faber, N., Jorna, R. Van Engelen, J.O. and Engelen, J.V. (2005) The sustainability of 'sustainability': A study into the conceptual foundations of the notion of 'sustainability'. *Journal of Environmental Assessment Policy and Management* 07(01), 1–33. DOI: 10.1142/S1464333205001955.

Fennell, D. (2015) *Ecotourism*, 4th edn. Routledge, London/UK.

Ferguson, P. (2014) The green economy agenda: business as usual or transformational discourse? *Environmental Politics* 24(1), 17–37. DOI: 10.1080/09644016.2014.919748.

Fletcher, R. (2011) Sustaining tourism, sustaining capitalism? The tourism industry's role in global capitalist expansion. *Tourism Geographies* 13(3), 443–461. DOI: 10.1080/14616688.2011.570372.

Fletcher, R., Murray Mas, I., Blanco Romero, A. and Blázquez-Salom, M. (2019) Tourism and degrowth: an emerging agenda for research and praxis. *Journal of Sustainable Tourism* 27(12), 1745–1763. DOI: 10.1080/09669582.2019.1679822.

Font, X. (2017) What meaning for sustainability? Creating tourism impacts in a slippery policy context. *Journal of Policy Research in Tourism, Leisure and Events* 9(2), 209–215. DOI: 10.1080/19407963.2016.1258514.

Forum anders reisen (2019) Kriterienkatalog des forum anders reisen e.V. Available at: https://forumandersreisen.de/fileadmin/user_upload/allgemeine_Infos_far/KK_neu_de_Stand_Juni2019.pdf (accessed 28 February 2020).

Foster, J.B. (2011) Capitalism and Degrowth: an impossibility theorem. *Monthly Review* 62(8), 26–33. DOI: 10.14452/MR-062-08-2011-01_2.

FUR – Forschungsgemeinschaft Urlaub und Reisen (2019) Reiseanalyse 2019. Selected first results of the 49th Reiseanalyse for the ITB 2019. Available at: https://reiseanalyse.de/wp-content/uploads/2019/03/RA2019_First-results_EN.pdf (accessed 3 March 2020).

Gascón, J. (2019) Tourism as a right: A "frivolous claim" against degrowth? *Journal of Sustainable Tourism* 27(12), 1825–1838.

Gössling, S. and Peeters, P. (2015) Assessing tourism's global environmental impact 1900–2050. *Journal of Sustainable Tourism* 23(5), 639–659.

Griffin, K. and Raj, R. (2017) The importance of religious tourism and pilgrimage: reflecting on definitions, motives and data. *International Journal of Religious Tourism and Pilgrimage* 5(3), ii–0.

Hall, C.M. (2009) Degrowing tourism: Décroissance, sustainable consumption and steady-state tourism. *Anatolia* 20(1), 46–61. DOI: 10.1080/13032917.2009.10518894.

Hall, C.M. (2010) Changing paradigms and global change: from sustainable to steady-state tourism. *Tourism Recreation Research* 35(2), 131–143. DOI: 10.1080/02508281.2010.11081629.

Hall, M.C., Scott, D. and Gössling, S. (2013) The primacy of climate change for sustainable international tourism. *Sustainable Development* 21(2), 112–121. DOI: 10.1002/sd.1562.

Heitmann, S., Robinson, P. and Povey, G. (2011) Slow food, slow cities and slow tourism. In: Robinson, P., Heitmann, S. and Dieke, P.U.C. (eds) *Research themes for tourism*. CABI Publications, Wallingford/UK, Cambridge, Massachusetts/USA, pp. 114–127.

Higgins-Desbiolles, F. (2018) Sustainable tourism: sustaining tourism or something more? *Tourism Management Perspectives* 25, 157–160. DOI: 10.1016/j.tmp.2017.11.017.

Higgins-Desbiolles, F., Carnicelli, S., Krolikowski, C., Wijesinghe, G. and Boluk, K. (2019) Degrowing tourism: rethinking tourism. *Journal of Sustainable Tourism* 27(12), 1926–1944. DOI: 10.1080/09669582.2019.1601732.

Jackson, T. (2009) *Prosperity without growth. Economics for a finite planet*. Earthscan, London/UK.

James, A., Ravichandran, S., Chuang, N.-K. and Bolden, E. (2017) Using lifestyle analysis to develop lodging packages for staycation travelers: an exploratory study. *Journal of Quality Assurance in Hospitality & Tourism* 18(4), 387–415. DOI: 10.1080/1528008X.2016.1250240.

Kallis, G. (2018) *Degrowth (The economy: Key ideas)*. Agenda Publishing, New York, New York/USA.

Kirstges, T.H. (2018) *Strukturanalyse des deutschen Reiseveranstaltermarktes 2018*. Dr. Kirstges Buch- und Musikverlag, Wilhelmshaven/Germany.

Lane, B. and Kastenholz, E. (2015) Rural tourism: the evolution of practice and research approaches – towards a new generation concept? *Journal of Sustainable Tourism* 23(8-9), 1133–1156. DOI: 10.1080/09669582.2015.1083997.

Latouche, S. (2009) *Farewell to growth*. Polity Press, Cambridge/UK, Malden, Massachusetts/USA.

Leslie, D. (2012) The responsible tourism debate. In: Leslie, D. (ed.) *Responsible Tourism: Concepts, Theory and Practice*. CAB International, Wallingford, UK, pp. 17–42.

Ma, M. and Hassink, R. (2013) An evolutionary perspective on tourism area development. *Annals of Tourism Research* 41, 89–109. DOI: 10.1016/j.annals.2012.12.004.

Mayring, P. (2015) *Qualitative Inhaltsanalyse. Grundlagen und Techniken*, 12th edn. Beltz, Weinheim/Germany.

Milano, C., Novelli, M. and Cheer, J.M. (2019) Overtourism and degrowth: a social movements perspective. *Journal of Sustainable Tourism* 27(12), 1857–1875. DOI: 10.1080/09669582.2019.1650054.

Moscardo, G. and Murphy, L. (2014) There is no such thing as sustainable tourism: Reconceptualizing tourism as a tool for sustainability. *Sustainability* 6(5), 2538–2561. DOI: 10.3390/su6052538.

Navarro-Jurado, E., Romero-Padilla, Y., Romero-Martínez, J.M., Serrano-Muñoz, E., Habegger, S. *et al.* (2019) Growth machines and social movements in mature tourist destinations Costa del Sol-Málaga. *Journal of Sustainable Tourism* 27(12), 1786–1803. DOI: 10.1080/09669582.2019.1677676.

Panzer-Krause, S. (2017) Un-locking unsustainable tourism destination paths: the role of voluntary compliance of tourism businesses with sustainability certification on the island of Rügen. *Zeitschrift für Wirtschaftsgeographie* 61(3-4), 174–190. DOI: 10.1515/zfw-2017-0013.

Panzer-Krause, S. (2019) Networking towards sustainable tourism: innovations between green growth and degrowth strategies. *Regional Studies* 53(7), 927–938. DOI: 10.1080/00343404.2018.1508873.

Panzer-Krause, S. (2020) The lost rural idyll? Tourists' attitudes towards sustainability and their influence on the production of rural space at a rural tourism hotspot in Northern Ireland. *Journal of Rural Studies* 80, 235–243. DOI: 10.1016/j.jrurstud.2020.09.013.

Park, H.Y. (2013) *Heritage tourism*. Routledge, London/UK, New York, New York/USA.

Paunović, I. and Jovanović, V. (2017) Implementation of sustainable tourism in the German Alps: a case study. *Sustainability* 9(2), 226–15. DOI: 10.3390/su9020226.

Powell, R.B. and Ham, S.H. (2008) Can ecotourism interpretation really lead to Pro-Conservation knowledge, attitudes and behaviour? Evidence from the Galapagos Islands. *Journal of Sustainable Tourism* 16(4), 467–489. DOI: 10.1080/09669580802154223.

Renkert, S.R. (2019) Community-owned tourism and degrowth: a case study in the Kichwa Añangu community. *Journal of Sustainable Tourism* 27(12), 1893–1908. DOI: 10.1080/09669582.2019.1660669.

Rifai, T. (2017) Tourism: Growth is not the enemy; it's how we manage it that counts. Available at: https://www.unwto.org/archive/global/press-release/2017-08-15/tourism-growth-not-enemy-it-s-how-we-manage-it-counts (accessed 3 March 2020).

Schumpeter, J. (2008 [1942]) *Capitalism, Socialism and Democracy*. Harper Perennial Modern Thought, New York, New York/USA.

Taş Gürsoy, İ. (2019) Beauty and the beast: a fairy tale of Tourismphobia. *Tourism Planning & Development* 16(4), 434–451. DOI: 10.1080/21568316.2019.1582086.

TourCert (2015) Durchführungsbestimmungen zur Zertifizierung von Reiseveranstaltern. Available at: https://www.tourcert.org/wp-content/uploads/2018/10/TourCert_Durchf%C3%BChrungsbestimmungen_RV_2015.pdf (accessed 28 February 2020).

TourCert (2018) Kriterienkatalog für Reiseveranstalter. TourCert. Available at: https://www.tourcert.org/wp-content/uploads/2019/04/TourCert_Kriterienkatalog_RV_2018-2.pdf (accessed 28 February 2020).

TourCert (2019a) Community. Available at: https://www.tourcert.org/community/ (accessed 27 December 2019).

TourCert (2019b) Reiseveranstalter: Zertifiziertes Nachhaltigkeitsmanagement. Available at: https://www.tourcert.org/wp-content/uploads/2019/11/Flyer-Reiseveranstalter_DE.pdf (accessed 27 December 2019).

UNWTO – United Nations World Tourism Organization (2019) World tourism highlights 2018. Available at: https://www.e-unwto.org/doi/pdf/ (accessed 3 March 2020).

Valdivielso, J. and Moranta, J. (2019) The social construction of the tourism degrowth discourse in the Balearic islands. *Journal of Sustainable Tourism* 27(12), 1876–1892. DOI: 10.1080/09669582.2019.1660670.

von Weizäcker, E.-U., Hargroves, K., Smith, M.H., Sesha, C. and Stasinopoulos, P. (2009) *Factor five. Transforming the global economy through 80% improvements in resource productivity*. Earthscan, London/UK.

Weaver, D.B. (2012) Organic, incremental and induced paths to sustainable mass tourism convergence. *Tourism Management* 33(5), 1030–1037. DOI: 10.1016/j.tourman.2011.08.011.

Appendix A: List of German TourCert certified tour operators (as at December 2019) and sustainability reports/audit reports

ID	Tour operator	Certified since	Website	Year of issue of sustainability report (sr) / audit report (ar)
1	a&e erlebnis:reisen	2010	www.ae-erlebnisreisen.de	2010 (sr), 2012 (ar), 2015 (sr)
2	Accept-Reisen	2010	www.accept-reisen.de	
3	Akwaba Afrika	2018	www.akwaba-travel.de	2018 (ar)
4	America Andina	2011	www.america-andina.de	2013 (sr), 2017 (sr)
5	Amitabha Reisen	2015	www.amitabha-reisen.de	
6	atambo tours	2010	www.atambo-tours.de	
7	AVENTERRA	2012	www.aventerra.de	
8	avenTOURa	2009	www.aventoura.de	2017 (sr)
9	Bayerisches Pilgerbüro	2015	www.pilgerreisen.de	2015 (sr), 2017 (sr)
10	Biketeam Radreisen	2011	www.biketeam-radreisen.de	2015 (ar)
11	biss Aktivreisen	2010	www.biss-reisen.de	2013 (sr), 2015 (sr), 2018 (sr)
12	Boundless	2018	www.boundless-reisen.de	
13	Chamäleon Reisen	2013	www.chamaeleon-reisen.de	2013 (ar), 2015 (sr), 2018 (sr)
14	CHINA BY BIKE	2010	www.china-by-bike.de	2015 (sr)
15	Cilentano	2010	www.cilentano.de	2018 (sr)
16	Club Aktiv	2013	www.club-aktiv.de	2012 (sr), 2015 (ar)
17	Daktari Travel	2011	www.daktaritravel.de	
18	Dein Marokko	2019	www.dein-marokko.de	2019 (ar)
19	drp Kulturtours	2015	www.drp-kulturtours.de	

ID	Tour operator	Certified since	Website	Year of issue of sustainability report (sr) / audit report (ar)
20	For Family Reisen	2009	www.familien-reisen.com	
21	France écotours	2010	www.france-ecotours.com	2018 (ar)
22	Gebeco	2011	www.gebeco.de	2019 (sr)
23	Geotoura	2011	www.geotoura.com	
24	Ghana Reisen	2014	www.kasapa.eu	
25	Hauser Exkursionen	2012	www.hauser-exkursionen.de	
26	Hummel Reiseideen	2018	www.hummel-reiseideen.de	
27	Island Erlebnisreisen	2017	www.islanderlebnis.de	2019 (sr)
28	Israel mal anders	2017	www.israelmalanders.de/	
29	Kolibri-Reisen	2010	www.kolibri-reisen.de	2015 (sr)
30	Laade Gartenreisen	2010	www.laade-gartenreisen.de	
31	lernen & helfen Sprachreisen	2011	www.lernenundhelfen.de	2011 (sr), 2013 (sr), 2016 (ar), 2019 (ar)
32	Lupe Reisen	2011	www.lupereisen.com	
33	MITourA	2017	www.mitoura.com	
34	Natours Reisen	2011	www.natours.de	2013 (sr)
35	Neue Wege Seminare & Reisen	2009	www.neuewege.com	2017 (sr)
36	nomad	2009	www.nomad-reisen.de	2019 (ar)
37	OCEANO MEERZEIT Reisen	2010	www.oceano-whalewatching.com , www.oceano-meerzeit.com	2014 (sr)
38	OLIMAR Reisen	2016	www.olimar.de	2016 (sr)
39	Papaya Tours	2011	www.papayatours.de	2016 (sr)
40	Picotours	2012	www.picotours.de	2014 (sr)
41	Radissimo Radreisen	2009	www.radissimo.de	2017 (ar)
42	REISEN MIT SINNEN	2009	www.reisenmitsinnen.de	2018 (sr)
43	ReNatour	2009	www.renatour.de	2014 (sr), 2017 (sr)
44	ROTH REISEN	2017	www.rothreisen.de	
45	Rucksack Reisen	2009	www.rucksack-reisen.de	2015 (ar), 2018 (ar)
46	SAPIO Kulinarische Entdeckungsreisen	2010	www.sapio.de	2015 (sr)
47	seabreeze.travel	2010	www.seabreeze.travel	2015 (sr)
48	Sento Wanderreisen	2014	www.sento-wanderreisen.de	
49	SKR Reisen	2011	www.skr.de	2018 (sr)
50	Solecu Tours	2010	www.solecu.de	
51	Tour Exquisit	2011	www.tourexquisit.de	
52	Travel To Life	2011	www.traveltolife.de	2011 (sr)
53	travel-to-nature	2009	www.traveltonature.de	2017 (ar)
54	Urlaub & Natur	2009	www.urlaubundnatur.de	
55	Ventura TRAVEL	2011	www.venturatravel.org	
56	Via Verde	2011	www.via-verde-reisen.de	2015 (sr)
57	Wikinger Reisen	2013	www.wikinger-reisen.de	2018 (sr)
58	Windbeutel Reisen	2018	www.windbeutel-reisen.de	
59	ZeitRäume	2014	www.zeitraeume-reisen.de	

7 The Kavatzas of Gavdos: Heterotopias Apart from Modern Societies

Pascal Mayer*

Université de Pau et des Pays de l'Adour, France

7.1 Introduction

Backpackers' guides about Crete generally give one or two pages to Gavdos, in which they always warn about frequent swell at sea that can suspend ferry services. Lonely Planet (2020) describes Gavdos as Europe's most southern point, a small island where there is little to do, except swim, walk and relax. The guide mentions 'a few rooms, tavernas and unspoiled beaches which attract campers, nudists and free spirits happy to trade the trappings of civilization for an unsullied nature experience'. In July and August, thousands of people camp on the beaches of Gavdos. This appeal deserves analysis.

There is very little data on Gavdos and its specific tourism. However, the qualitative primary research conducted by Andriotis (2013) constitutes a basis. Over two summers on Gavdos in 2010 and in 2011 he studied a specific group of travellers involved in what he calls an *antinomian counterculture* (in opposition with norms). According to him, economic crisis in Greece as in many European countries brought many young people to rebel against their own culture. He notes (p. 12): 'Like 60s and 70s hippies, who in protest against industrial society used to sleep in the caves of Matala [Crete], the antinomian group has for a while now felt alienated in society and chosen to turn their back on the system and its values. Thus, the crack in the traditional values of the puritan world gave voice to the hippies who prepared the ground for the birth of antinomianism.' Andriotis called this group 'seasonal antinomians' but he also mentions two other groups – not studied – on Gavdos' beaches: first, people attracted by free camping opportunities rather than the alternative lifestyle, and second a core of permanent antinomians.

In fact, other groups (or subgroups) can be identified on the beaches of Gavdos but the seasonal antinomians described by Andriotis (2013) constitute

*pac.mayer@laposte.net

DOI: 10.1079/9781789245073.0007

the reference group that gives its specificity to the island. The geographical location and the social situation of Gavdos were factors that prevented the emergence of a standard tourist offer and promoted a radical degrowth tourism which has become something of the norm, of which *kavatzas* – the shelters installed under little cedar trees on the beaches – are probably the most emblematic elements. Newcomers to Gavdos cannot isolate themselves from this reality and they have to share the local way of life on the beaches or run away. A stay on Gavdos (even in standard accommodation conditions) is more often referred to as an 'experience' than a 'visit'. The purpose of this study is therefore to try to identify the commonalities that link the different groups of individuals who frequent Gavdos and to define in which way this island, its beaches and its kavatzas constitute privileged spaces to escape the constraints of modern societies. This is a complex subject which needs a cross-disciplinary approach.

7.2 Methodology

This study was conducted as part of an International Human Ecology Certificate issued by the University of Pau (France). Human ecology is interested in the complex interactions between ecosystems and social systems through a transdisciplinary approach (Morin, 1979; Crognier, 1994; Jacquard, 2004). Gavdos has a very specific ecosystem and basic social structure that have allowed in the past the development of an episodic antinomian community that today colours the entire life of the island. A located utopia, something of a heterotopia (as described by Foucault, 2009) has been created where rules are 'other' and allow everybody to feel free within nature in an idealized primordial way of life in which everyone adjusts to his/her personal opinions, constraints and limits.

To explore the reasons why Gavdos exerts such a power of attraction, extensive literature research and qualitative methods were used. The literature research covered a wide spectrum of the ideas of naturalism, antinomianism, degrowth and heterotopia. Furthermore, two naturalistic data collection techniques were adopted: in-depth participant observation and inquiries (Van Campenhoudt *et al.*, 2017). Two distinct populations were interviewed, locals (nine in total) and people frequenting the beaches (27 in total) during three periods: September 2018, April/May 2019 and August/September 2019.

For natives (or their children now engaged in touristic activities), unstructured interviews were held in the professional context of these people, during quiet moments. After a quick presentation of the objectives of the study (In what way are Gavdos, its beaches and its kavatzas special spaces of tourism or life very similar to achieved utopias?), the interlocutors were invited to tell a story of their lives, to specify the choices that had made them stay on or return to Gavdos and then to express their visions or wishes for the future.

For tourists (or people living permanently in kavatzas), the interviews took place on the beaches. Care was taken to visit all the beaches of Gavdos at different times of day, and in different contexts (by the sea, near freshwater springs or near a kavatza) so that statistically a great number of profiles could be identified. The discussions were generally held in English, sometimes in French. They were unrecorded because many of the interviewees feared inappropriate use through social networks or a publication in the press that could increase tourism on the island. The average age of respondents was slightly less than 40 (the youngest was 21 and the oldest was 66). Of the 27 interviews conducted on the beaches, three concerned young people travelling in a couple, two concerned young people travelling in groups (a group of two and a group of three) and two concerned women travelling alone with their children (one 10 years old and the other 16 years old). The bulk of the interviews (20) therefore relate to people travelling alone. There are relatively few families and children on the beaches of Gavdos, except in Sarakiniko where most of the island's conventional accommodation is concentrated. In terms of gender, 60% of the interviewees were men. In terms of nationality, 60% were Greek; the rest were Italians, Germans, French, Swiss, English, Canadian and Portuguese. The interviews were semi-structured. This means that ten topics had been selected to be addressed but discussions were informal, and everyone was free to debate other themes. Conversation was focused on the following points: context of the first visit to Gavdos, description of a usual stay on Gavdos, type of accommodation, everyday organization, favourite activities, social life on the beach, advice to enjoy a stay on Gavdos, specificities of Gavdos and plans for the future.

In a final stage, data were clustered around themes that enabled a flexible classification of people frequenting the beaches of Gavdos – larger than the one given by Andriotis (2013). It emerged that at least one of the following three arguments was systematically raised during the interviews, with a more or less strong intensity, to justify a stay on Gavdos: detachment from material and social constraints, harmony with nature or significant personal step. By combining these themes with the type of accommodation (wild camping or conventional room) and the length of stay (short, long or permanent), five distinct profiles finally emerged with several sub-profiles: seasonal community members, seasonal asocial individuals, permanent sedentary marginals, occasional degrowth travellers and involved conventional tourists.

Spaces of nature, islands and beaches have already been studied as heterotopias (Bigell, 2007; Rieucau and Lageiste, 2008; Andriotis, 2013; Melay, 2019). The kavatzas seem to bring something more to beaches as heterotopias. These basic shelters, quite specific to Gavdos by their number, facilitate a flashback to childhood and to the origins of humanity. They unlock the imagination and allow attempts to respond to the problems of our modern societies: materialism, overexploitation, crushing of the individual. Thus, it appeared that considering the kavatzas as pieces of heterotopia, analysing the functions they perform, was useful to illustrate the very special appeal of Gavdos to a wide variety of individuals.

7.3 Gavdos Kavatzas: Original Natural Habitats

7.3.1 The island of Gavdos, a margin

Gavdos is the only small inhabited island south of a large island, Crete, 40 km away. Its surface area is only 33 km^2 – about 10 km long and 5 km wide (Stefanakis, 2003). Reaching Gavdos is not easy. From Heraklion or Chania on the north coast of Crete, where tourists usually land, you have to take a bus to reach Paleochora or Chora Sfakion on the south coast and then a ferry that runs only on certain days of the week and stays docked as soon as the wind exceeds 5 Beaufort, which is common, even in summer. From whatever European capital you leave, you cannot reach Karave, the harbour of Gavdos, in less than two or three days of travel and the swell can delay things further by several days.

The geographical object 'Gavdos' has something very special: it is the edge of a world, in both physical and socioeconomic terms. From a geological point of view, Gavdos is the first ledge – before the Cretan coast – of the microplate of the Aegean Sea under which the African plate sinks. Its south coast is therefore elevated, made up of steep cliffs that rise abruptly to more than 300 metres above sea level (Galanakis, 2009). The altitude decreases slowly down to the north and the sandy beaches. Because of this relief, the island receives some precipitation and some of the beaches have fresh water sources that are very important for free camping. Today, the island is mostly covered with a clear forest of pines and junipers. First land north of Libya, it is a particularly valuable biotope for birds in their migrations.

In the past, fresh water supported agricultural development that sustained a population that reached 1000 people. An exodus was significant from the 19th century onwards and the low point was reached at the end of the 20th century with less than 50 people living on the island during winter. There is no sign of agricultural renewal and few cultivated fields left (partly because of more frequent droughts), except those tended by an esoteric sect installed in the heights of the island since 1997 on land donated by the pope's mother at the time. They are known as the 'Immortals' or the 'Russians' (because the founders were Russian physicists and one of them was irradiated at the Chernobyl site) who chose to abandon their careers, live a 'real life' and study philosophy far away from big cities. They conduct incomprehensible spiritual research within the 'Pythagorean Institute of Philosophical Studies for the Immortality of Man'. But they work hard, cultivate the land and help local society. 'They are good people!' remarked a 45-year-old studio renter in Sarakiniko. 'They are serving others, especially in repairing cars and machines.'

Fishing is the occupation of only three people, two of whom also run taverns with their families. From an economic point of view, Gavdos is one of the least developed areas in all Greece. Random ferry services explain the initial lack of tourism investment. A power grid was not installed until 1990. The social rules on the island are far removed from the standards of the rest of Europe. There is one teacher for three pupils and one policeman. One of his duties is to drive pupils to school. Gavdos has nothing to do with the smart islands of the Cyclades, north of Crete. Here, no lovely little harbour and no freshly painted white houses. No petrol station, no bank, no shop and no nightclub of course, and only one bakery in Kastri, the unattractive village in the centre of the island.

Tourism is today the main activity. Gavdos offers about 120 beds and about 25 taverns, half of them in Sarakiniko. Most people operating in tourism are children or grandchildren of island's natives but they were not born on Gavdos and have spent a significant part of their lives elsewhere. They generally don't stay over the winter. They took advantage of low budget tourism that emerged in the 1980s by offering adapted services. A woman in her early fifties who was one of the first room renters in Sarakiniko said: 'I lived in Chania [Crete] until I was 12 years old. Then my father who was born on Gavdos came back to the island with the whole family in order to work. I had been staying for five years in the upper part of the island and then I came here to Sarakiniko to begin the work I do now. In winter I go back to Chania where my children live. I'm here from Easter until the end of October. There are not many establishments open that long except in Kastri, where there is now a year-round establishment. Some rich people from Athens may come by helicopter in winter to enjoy an absolute quietness!'

These tourist establishments employ staff, sometimes local, sometimes Greek tourists looking to earn some money to finance their stay, but more often foreign workers, many of whom are deeply unsatisfied. An Albanian in his forties who was leaving the island upset, complained: 'In Greece and especially in Gavdos, all those who work are Albanians, Romanians, Bulgarians or Georgians. The Greeks, they watch working, they eat, they drink. In Gavdos we are paid even less than in Crete. We aren't housed and can't afford housing. We have to camp on the grounds of the tavern that employs us, in a hidden corner, or on the beach. We manage for the rest: food, washing, etc. But everything found in the minimarkets here costs twice the price of Paleochora [Crete]. If we don't accept to work, there will be hippies to accept!'

7.3.2 The beaches of Gavdos: specific ecological and social systems

There are six beaches with water supply and free camping opportunities (Fig. 7.1). Each of these has its own character, in terms of orientation, sunshine, wind, view and distance from amenities. On the north coast of Gavdos, about 1.5 km from the little harbour of Karave, lies Sarakiniko, the biggest and most equipped beach from a tourist point of view. Further to the west, about 3 km from Sarakiniko, lies Agios Ioannis, the most emblematic beach of Gavdos. Near the edge of the north coast, 1 km from Agios Ioannis, lies Lavrakas, which is equipped with a well and a pump directly on the sand. From any of these three beaches, one can directly contemplate the high silhouette of Crete. Oriented to the west lies Stravolimni, only 0.5 km

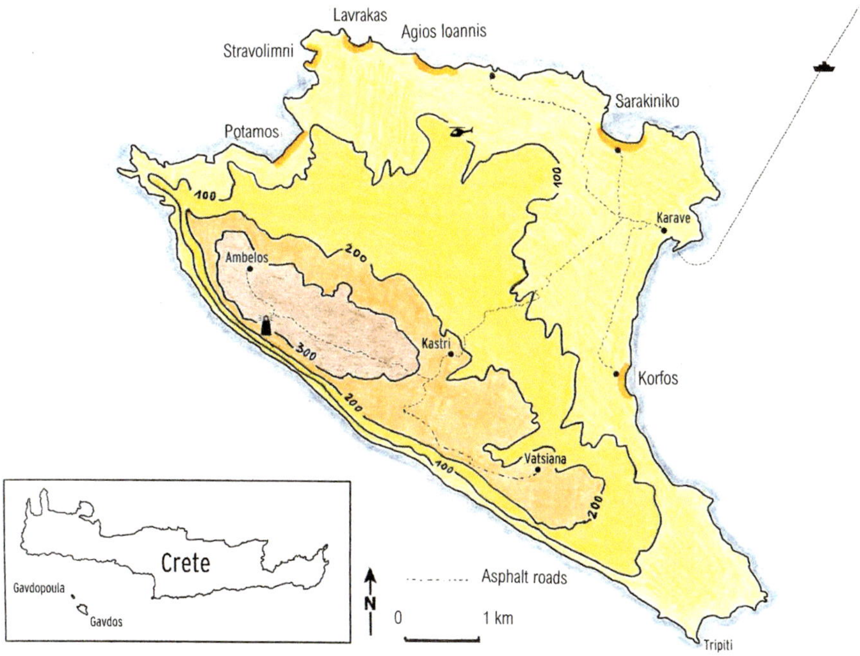

Fig. 7.1. Gavdos and its beaches.

from Lavrakas. South from Stravolimni, and difficult to reach from any side, lies Potamos. From Stravolimni and Potamos, one can see the silhouette of Gavdopoula, the little uninhabited island of Gavdos. On the east coast, about 2 km south of Karave, lies Korfos, well sheltered from the wind.

The whole island is classified as Natura 2000 and the coastal dunes with *Juniperus* species are considered to be of the highest ecological and aesthetic value. According to the EU Habitat Directive, they are a 'priority habitat', in danger of disappearing (Galanakis, 2009). That clearly means that everything that has been built in Sarakiniko (studios and taverns) is technically illegal.

According to one of the fishermen, now in his sixties, who ran the island's first tavern, it all started in 1978. 'It was very different from today: only 20 to 30 people in tents on Sarakiniko beach where there was nothing.' The fisherman's tavern was a small remnant of a communist deportee camp set up on this beach by the dictatorship of General Metaxas shortly before World War II. This origin is confirmed by one of the first settlers, a Greek from Athens in his sixties now permanently residing in an isolated kavatza:

> 'When I was young, I loved travelling. I wasn't a hippie but I wanted to see how other people were like. I've done a lot of jobs. I used to hang out with communists, I knew Mikis Theodorakis. I first came to Gavdos in 1978. Even between Karave and Kastri it was only a track with donkeys: no car, no electricity. There were only three *kafeneio* on the island, in Karave, Kastri and Vatsiana. We were the first in Sarakiniko, 30 or 40 of us living there all summer. In 1979, I stayed all year round, next to the house of the exiles. The

> following years, I worked in Athens or Crete in the winter and I came to spend the summer on Gavdos, on Agios Ioannis beach or Sarakiniko beach. I have been staying permanently here for six years.'

A second wave followed in the 1990s, more anarchist than leftist, as explained by a woman in her late forties, accompanied by her 10-year-old son: 'I came for the first time to Gavdos in September 1990. I had heard about it in Athens, in the Exarchia district that I frequented at the time, where punks and anarchists gathered. When we got off the boat, tractors took us from Karave to Kastri or from Karave to Sarakiniko. There were two or three taverns in Sarakiniko, but only one opened in September. We stayed a little away from the thirty-somethings, the group before us, more politically oriented than we were.'

A decade later, the seasonal antinomians studied by Andriotis (2013), led by the Greek crisis, constituted a third wave, in continuity with the first two. A division is observed today as pointed out by a student on a two-week holiday in Potamos: 'The beach of Agios Ioannis has become very fashionable among some young people in Athens. They do not say – this year I go to Gavdos, but this year I go to Aï Yannis [a hip diminutive for Agios Ioannis]!'

7.3.3 The kavatzas of Gavdos: ephemeral hiding places

Two juniper species grow on the beaches of Gavdos: *Juniperus phoenica* and *Juniperus macrocarpa*. The first one is a spreading shrub but the second one, the most common on the beaches, can be a small tree, offering shade and shelter thanks to very dense foliage. Strictly protected from an ecological point of view, it is locally called 'cedar'. Fortunately, the root system of this tree is very deep (to withstand the wind and find water) and is therefore insensitive to permanent trampling on the beaches.

Scattered shelters utilizing cedars are called kavatzas, a slang word that etymologically means 'small cellar' but must be translated instead to 'hiding place', 'fall back' or 'plan B'. Kavatzas can also be installed under a pine or even a filao whose quality of shelter is not so good. They are not specific to Gavdos but nowhere in Greece are kavatzas so numerous, and nowhere else in Greece does the dominant mode of tourism rely on them. These shelters are well hidden. The layout usually consists of a small tent securely attached to the tree or to rocks in the soil. One or several tarpaulins can add protection against the wind. But you can also simply put down your backpack and your sleeping bag (or stretch your hammock) under the shade of the tree (Fig. 7.2).

Sometimes, real huts have been built from stone and pieces of wood by those who spend very long periods of the year on the island or stay over the winter. These huts (maybe 30 in all) are curiosities and a source of pride for those who built them. But the huts do not correspond to the spirit of the beaches of Gavdos according to a regular from the anarchist wave: 'I do not like huts. This suggests to those who built them that they

Fig. 7.2. A basic kavatza in Lavrakas: a cedar, a few stones and a tarpaulin. In the background, several kavatzas are hidden among the trees.

have rights. And some stay there for a couple of years and then suddenly disappear, leaving everything behind and the place becomes a dump. I prefer what is ephemeral: one summer under a tree, next summer under another tree!' Very often, without forming a hut, some structural elements have been brought in: dry stone walls, wood benches, types of storage. In Agios Ioannis, Lavrakas and Potamos, there are kavatzas-kafenion. These are kavatzas very close to the sea, with benches, tables and floated wooden shelves where everyone can come and chat, drink coffee or raki. No one has the right to settle there. The regulars know how to inform the newcomers unaware of this rule.

7.4 Gavdos Kavatzas: Symbols of Antinomianism

The first group of leftists that camped in Sarakiniko in 1978 was probably homogeneous. Now different profiles can be found on the beaches even though something still floats in the air of Gavdos, related to antinomianism. Below is a flexible classification of kavatzas users.

7.4.1 Seasonal community members

Andriotis (2013) explored in Agios Ioannis and Lavrakas antinomians' specific travel patterns, motivations, activities, norms, beliefs, appearance and forms of social interaction. He called this group of travellers 'seasonal antinomians', considering that they reject established norms, behaviours, values, beliefs and consumption patterns of affluent society. As antinomianism colours the entire life of Gavdos, we call here this group 'seasonal community members'. It includes 'leftist and anarchist veterans' and it is regularly augmented by young Greeks with strong convictions about freedom and also preoccupations reflecting the social tensions of the moment – anti-capitalism during the Greek crisis when Andriotis (2013) conducted his 26 interviews; learning from nature in 2019. Andriotis identified, rather, young people taking flexible trips each year to Gavdos for extended periods of time, longer than the brief vacations of mass tourists, travelling lightly with a backpack, recognizable by their long, dishevelled hair, either naked or wearing beach wraps. Most of the travellers interviewed by Andriotis linked their enduring involvement within the antinomian counterculture with societal alienation and economic crisis that had created for some of them a transitional phase in their life, mainly associated with career disruption. They found that they had a chance to experience in Gavdos adventures denied to them in everyday life. Living primitively seems to bring both greater freedom and a higher level of consciousness through interaction with surrounding nature.

According to Andriotis, this group has little interest in meeting locals and learning about their culture. In practice, the main form of socializing with locals seems limited to commercial transactions, either when working or when buying. Targeted interaction – spending time with like-minded travellers – is therefore at the heart of this group's journey. The strong cohesion and collective spirit of the community in an isolated place, far away from civilization, has created a kind of shared territory with geographical boundaries. This interaction is enhanced through social nudity and drug use (essentially joints). Smoking in a group is a kind of secular sacrament, the same one which served as a collective bond among hippies. Time spent in conversations, in exchanges of ideas that could change the world (or not) without imposing a barrier is important. Newcomers, provided they meet certain criteria, are always welcome.

In the consumerist world of the 21st century, antinomians try to live free from money, material possessions, tickets, reservations or any of the attributes of mass tourism. On the beaches, there is no electricity, no running water, no light pollution at night. The absence of material demands allows extended stays since it's possible to live on Gavdos for less than 300 euros per month. As financing methods, Andriotis (2013) lists financial support of parents, short-term work done before the stay, handicrafts trade or casual jobs in the local economy.

The reflections of a young Greek in his thirties, encountered in April 2019 in Agios Yoannis, illustrate what Andriotis describes, but with more nature-oriented preoccupations: 'I arrived in February. I think I'll stay for several months, maybe spend the winter. Life is more valuable when you do things directly in nature because you give more time to the things that matter most: fetching water, collecting wood (never cedar wood), making tea, cooking, doing the dishes with sand and sometimes ash, swimming, collecting salt from the hollows in the rocks , doing nothing ... I also make handicrafts from abandoned things, usually wood. I don't recycle, I upcycle … I don't have any particular philosophy. I do not belong to any religion or ideology. The current crisis is not economic, it is in our minds. Nature knows nothing about money. Nature educates. We must find a way to be in harmony with the planet. There is no need for rules if you are conscious and aware. Laws exist only for those who are not aware of what they are doing. For me, life is better without any rule. I camp in the most beautiful kavatza on the island. Friends of mine suggested that I could settle there. I will let [leave] the place exactly as I found it when I arrived.'

The importance of this group is highlighted by the testimony of a young Portuguese man in his thirties who had come to Gavdos every year for three years: 'It's a good thing that the regulars of Gavdos are well spread out on the beaches. When a problem arises, they are here to manage it. They intervene calmly and give advice. I saw them organizing the fight against a brush fire that a moron of August had provoked. It was very effective!'

7.4.2 Seasonal asocial individuals

Individuals in this group differ from the seasonal community antinomians studied by Andriotis (2013) in two respects:

- It is more because of a crisis in their personal lives rather than due to political or social awareness that the option of long stays on Gavdos became attractive. Their political conscience is weaker than that of community antinomians.
- For this reason, they are less inclined to be part of a community. They have wounds to heal and have a certain mistrust of others. They generally cut off relationships with family and old friends.

These individuals share with seasonal community antinomians the same rejection of materialism and the same sense of harmony in living naked with nature. They appreciate not being solicited by their neighbours (people in the kavatzas around them) and socialize only when they wish to.

A man in his late forties, from Germany, encountered in September 2018, explained: 'In 2015 I divorced, I lost my job, everything went wrong. I took a break for a year, I spent half a year in Paleochora [Crete] that I already knew and half a year here in Gavdos from April to October, first in Sarakiniko and then here in Potamos. I came back in 2016, not in 2017 but again this year. I go back to Munich for winter.' The same man said, in April 2019, arriving in Sarakiniko from Crete where he had spent the winter: 'That's it, I threw my papers and my phone into the sea. I'm doomed to stay here where I can experience how I can live with nothing. When you are in such natural surroundings, you can get energy from the sand, from the water, you can live without eating, you understand that you can be a God.'

Seasonal asocial individuals have no specific favourite beach and they can move from one to another during the time they stay on Gavdos. In summer, they just keep away from Sarakiniko, Agios Ioannis or Lavrakas because of the crowd. What is important to them is a kavatza that is rather isolated and quiet. Some of them (like the man mentioned above) have in a way to become residents and marginals.

7.4.3 Permanent sedentary marginals

There are about 20 people living permanently on the beaches of Gavdos, most of them at Lavrakas and the rest (during winter 2018–2019) at Agios Ioannis (three), Sarakiniko (three), Stravolimni (two) and Potamos (one). For some, this represents the most radical form of antinomianism: the way to go to the end of their rejection of modern, materialistic, consumerist and alienating society. For others, it's just pragmatism. Couples are rare. About half the permanent marginals have an activity, either temporary (related to tourism) or permanent (municipal jobs).

Three sub-profiles have been identified:

- Those who have spent many summers on the island within the antinomian community and finally decided to stay, considering that this was the best place to live in accordance with their ideals. They remain attached to their convictions and are pleased to find some of their former travelling and struggling companions when these come to spend a few more weeks on Gavdos. A Greek man in his sixties in Potamos told us: 'I don't know if I'm a communist or an anarchist but I know I'm an environmentalist! How not to be when you live here! I respect cedars and I make others (these horrible tourists!) respect them. I only burn lentisk wood. My kavatza, I chose it in nature, well sheltered from the wind, quiet, with a nice view, near the spring of which I drink water. I built it with the help of a friend six years ago. There are two small rooms, the living room with the fireplace and the bedroom. We installed a wooden floor and I installed real panes. I have solar panels; so I can use my computer and my phone. Every day I walk on the beach. I pick up what can be useful, like pieces of wood and collect the waste in a corner so that someone evacuates it with a canoe. I'm happy with my life. I've done a lot of things, I drank the best wine of Bordeaux. I don't know what I'm going to do later. Maybe I'll go back to Athens.'
- Those who are living out a personal process and have found, by chance sometimes, a satisfying lifestyle on a Gavdos beach. They have no political or philosophical convictions but don't want to play a social role any more in common life; like this Greek man, in his fifties installed in a kavatza in Lavrakas: 'When I turned 40, I realized that I was tired of everything: motorcycles, races, diving, women, I had done everything. I had been an engineer and a DJ too. I was getting old, the nights had changed. I settled on a beach near Athens. I heard about Gavdos from friends. They told me: you're from there! I came here in 2015, four years ago with nothing. I chose this kavatza because it overlooks the beach. I can see the rising sun and the setting sun and I am protected from the wind by a few bushes in the front. I built everything myself around a tent, with wood. Here the living conditions are good all year round. I can spend entire days looking at the sea. I don't read, I don't meditate, I don't do yoga, I don't walk. I have a canoe. I like to play with dolphins. This kavatza, it's the only place for me, my last place. I call it my place.'

- Those who decided to stay and work and have no other housing possibility than kavatzas, like this Greek man in his late thirties: 'Before the Greek crisis, I came here only on holiday. Now I work here and when working on this island it is convenient to have a tent and a car in Sarakiniko. I've been using the same kavatza for two years. Most important is the shade it provides. And for me, it's also important to be able to carry things easily from my car. I settled away from the sea as there are a lot of people in the summer (and I need to be quiet) and close to the road. I have a big solar panel, a computer, a big stove for cooking. It's really nice to live here. I know a lot of people on the island, especially people from here. I have a lot of hobbies, projects. Sometimes in the summer I move to Agios Ioannis to enjoy the atmosphere, to have fun. I don't have a particular philosophy of life. I can't say I'm an environmentalist when I'm using plastic!'

Varvarousis (2012), who studied the community in Lavrakas, notes an interesting feature about dwelling in what he calls an 'eco community': 'Unlike what usually happens in the city, the houses [here kavatzas turned into huts] are not designed in order for someone to live there, but they are shaped throughout someone's living there. This dynamic process results in the houses serving better, reflecting the views, the emotions and the psychology of their owners.'

7.4.4 Occasional degrowth travellers

There are now great numbers of visitors to Gavdos in July and August. Here are some sub-profiles:

- Young people, mainly Greeks, often students from big cities, who appreciate being able to spend very inexpensive holidays, without constraints of any kind in an environment that looks like paradise. Spending a few days or weeks on Gavdos is a kind of initiation rite, a form of militant affirmation of the doubts they have about the society they will have to face during their professional lives. Many of them travel as a couple and move from one beach to another during their stay. The beach of Agios Ioannis, however, is an obligatory crossing point. A Greek student in architecture wearing dreadlocks, in Potamos with his girlfriend, remarked: 'What we most like here is simplicity and absolute freedom. We walk a lot, we enjoy the beach, the friends, the fires, the food together. I draw, I carve wood.'
- Older individuals, of various European nationalities, engaged in a professional life and integrated into modern society, who have heard about Gavdos and come to test their ability to be alone with nature, to live a degrowth period in their lives. They avoid July and August and Agios Ioannis. They are observant, attentive and curious, both to the world around them and to their own sensations, like this Italian woman in her fifties met in April 2019 in Stravolimni: 'I have a fulfilling job and

children. I have some money and I am free. I don't have any particular philosophy but I like what is essential. Friends told me about Gavdos. Last year I stayed in a bungalow in Korfos. And this year for Easter, I chose to camp in Stravolimni because of the sunset. A friend of mine who was leaving told me about this kavatza that she had arranged a little. This is the first time I use a hammock. Living in a kavatza is something paranormal for me but I think I will do it again. I think this type of stay helps to stand normal life afterwards.'

- Tourists of all ages and nationalities, who are attracted by camping and life within nature. They find authenticity in their practice. They sometimes have a car or a scooter. They can be found on all the beaches of Gavdos depending upon their affinities. A French man in his late sixties: 'I've been coming here for 20 years. I always camp in Korfos. That's the best place for me. It's calm and there is no wind. I come by car from France and I stay alone for one or two months, depending on the holidays of my grandchildren in France. When I leave, I [store] my tent and all my stuff in a room of the nearby tavern. I drink water from the spring and I have dinner in the tavern only one day out of two.'

7.4.5 Involved conventional tourists

Even people with accommodation may use vacant kavatzas to shelter from the sun on the beach. There is indeed no umbrella and sunbed rental company on any beach of Gavdos. There is thus a closeness between 'free campers' and conventional tourists.

Some conventional tourists may have had an experience of kavatzas in the past and may have given up for comfort reasons (ageing people) or tranquility (single women). A French woman in her sixties, an ex-lawyer, who regularly stays at an establishment in Korfos: 'I first came to Gavdos in 1996. I met some very original people, very funny. I was renting a small room in Sarakiniko where I came back every year. I had never camped in my life and the fifth year, I wanted to do like my friends. I set up a tent in Sarakiniko but I couldn't rest. It was too hot, there was music all night long and the guy from the kavatza next door was staring at me! I gave up.' Like this lady, there is a high percentage of conventional tourists considering (dreaming) that one day they will settle in a kavatza, or at least give it a try. More simply, conventional tourists who frequent Gavdos have sympathy or at least a very high tolerance for those who camp on the beaches.

Most conventional tourists share many values with those who live in kavatzas, as explained by the same lady: a) distance from the modern world, lack of market solicitations, simplicity of living: 'Here there are no clothes stores, but very nice stones to look at'; b) direct contact with nature: 'I love nature. Here I feel in osmosis with it, I have animal sensations, ecstasy when I walk alone on a path surrounded by scents. Gavdos brings a man back to his true nature, before civilization, comfort and consumption'; c) freedom to be and act: 'Here you understand that you don't need clothes!'

7.5 Gavdos Kavatzas: Crystals of Heterotopia

Heterotopia (from the Greek *heteros* and *topos*, place) is a concept elaborated by the philosopher Michel Foucault to describe certain cultural, institutional and discursive spaces that are somehow 'other': disturbing, intense, incompatible, contradictory or transforming. He evoked this concept for the first time in a radio broadcast of 1966 (Foucault, 2009) entitled 'Other spaces' in which he defined heterotopia as a physical location of utopia or as a parallel space. These spaces, where rules are different, are concrete and host the imagination or are used to exclude or to host specific activities or people.Foucault identifies six principles for a systematic characterization of heterotopias:

- Heterotopias exist in any culture, in various forms, be they primitive or modern societies. A 'crisis heterotopia' is a separate space like a boarding school or a motel room where activities like coming of age or a honeymoon take place out of sight. They are reserved for individuals who are, due to society and to the human environment in which they live, in a state of crisis. They are less common in modern societies and get replaced by heterotopias of deviation which are institutions where we place individuals whose behaviour is outside the norm (hospitals, asylums, prisons, rest homes, cemeteries).
- The same heterotopia can see its functioning change over time.
- A heterotopia can juxtapose in one place several spaces themselves incompatible in real space. A garden can be a heterotopia, a microcosm of different environments, with plants from around the world.
- Within a heterotopia there is a heterochronia, i.e. a break with real time. *Heterotopias of time,* such as museums, enclose in one place objects from all times and styles. They exist in time but also exist outside of time.
- Heterotopias can open and close, which isolates them, makes them accessible and penetrable. Heterotopias of ritual or purification are spaces that are isolated and penetrable yet not freely accessible, like a public place. Either entering the heterotopia is compulsory, like entering a prison, or entry requires special rituals or gestures, as in a sauna or a *hammam*.
- Heterotopias have a function compared to other spaces in society: they are either spaces of illusion or spaces of perfection.

Postmodernist geographers (Harvey, 1989; Soja, 1989) have often used the concept of 'located utopia' or 'heterotopia' to help understand the contemporary emergence of 'difference' and 'identity' as central issues in modern multicultural societies, often in relation to ethnicity and gender. In our context, we have to consider that the question of difference and identity is somehow related to the materialism issue: am I myself through all my belongings or am I myself through every sensation I can feel, surrounded by nature, in kavatzas that are ephemeral shelters?

Johnson (2013) explored the ongoing fascination with Foucault's brief and rather sketchy idea of heterotopia. He notes that despite (or perhaps because of) the fragmentary and elusive quality of the idea, the concept of heterotopia has generated a host of conflicting, heterogeneous and almost mischievous interpretations and research across a range of disciplines (art and architecture, literacy studies, geography, sociology and urban studies). He considers that a key aspect of heterotopias is sometimes missing in research papers: their imaginative quality. He cites a striking passage that appears only in the first radio broadcast by Foucault in 1966: 'These counter-spaces, these locally realized utopias, are well recognized by children. Certainly, it's the bottom of the garden; or even more, it's the Indian tent erected in the middle of the attic; or still, it's Thursday afternoons on their parent's bed. It is on that bed where they discover the ocean, as they can swim between the covers, and the bed is also the sky, or they can bounce on the springs; it's the forest as they can hide there; or still, it's night as they can become ghosts between the sheets and, finally, it's the delight, as their parents come home, as they may be scolded.' Each kavatza clearly has something to do with a parent's bed. They are spatio-temporal units that interrupt the stream of everyday experiences, opening spaces of rest, refuge and imaginative play.

If kavatzas can be considered as pure heterotopias– natural spaces– islands and beaches have also been considered as heterotopias by different authors. Bigell (2007) considers that Foucault's concept of heterotopia provides a useful tool for understanding the social function of natural spaces. For him it is the relative openness for individual signification that defines a heterotopian space and natural spaces are predestined heterotopias because of their distance from cultural centres, scarcity of human population and lack of infrastructure – all of which contribute to a lessening of social control. For Doyon-Gosselin and Bélanger (2013), islands are heterotopic in the way they are absolutely real, but they touch us with dreams or metaphors. They cite Deleuze (1953): 'Dreaming of the islands, with anguish or with joy, whatever, it is dreaming that we separate, that we are already separated, far from the continents, that we are alone and lost, or it is dreaming that we start from nothing, that we recreate, that we start again.' For Melay (2019), islands, because of their isolation, often mean untouched and non-perverted spaces, apart from time and space. For Rieucau and Lageiste (2008), the beach is an untamed space of wilderness, an atypical space that has nothing to do with the host society, a place both constant and evanescent where everyone can escape. The body, as an object of pleasure, maintains with the beach a privileged relationship with the erotic dimension of sand and the foetal dimension of water.

7.5.1 Gavdos and its kavatzas, heterotopias of crisis or deviation

For Bigell (2007), natural spaces facilitate the unsettling and resettling of the character, considering that some existential crises have combined the physical with the spiritual and led to a reassessment of life. We have seen that personal crises, sometimes amplified by economic crisis, are a motivating factor to spend long periods of time on Gavdos for seasonal asocial individuals and for some of the permanent sedentary marginals when, sometimes, despite a real social success, they suddenly realize that they lack something, that they need to change their life radically. For those who have known this kind of crisis and have decided to settle, living in a kavatza on Gavdos is always considered an achievement, a final and luminous step. 'I couldn't live elsewhere!'

Antinomianism, behaviours outside the norm, is part of the spirit of Gavdos and it colours all social life. It includes rejection of bourgeois authority, social constraints, materialism, capitalism and consumer society. It leads to devaluation of work and distance from family. Beaches are typical heterotopias of deviation according to Rieucau and Lageiste (2008), removed places, where certain practices are made possible: idleness, nudity, promiscuity, drug use and drinking are the ultimate stage of deviation. And living in a kavatza means reinvesting in nature in order to feel free and distant from modern society.

7.5.2 Gavdos and its kavatzas, heterotopias in constant evolution

The prison camp for communists built on Sarakiniko beach before World War II was the first heterotopia on the island, at a time when the island was not so different from other small Greek islands in terms of development. Then, after the war, when Gavdos almost turned into a desert island because of strong agricultural decline, different waves of voluntary seasonal exiles followed. They were politically or philosophically oriented: communists in the 1980s, anarchists in the 1990s, antinomians more recently ... such visitors have increased a lot and the profiles of people coming to Gavdos are now less homogeneous. Their preoccupation seems to be less 'changing the world' than 'finding ways to feel free and self-conscious through a return to one's roots and nature'. A new phenomenon illustrates this change: groups (about ten in 2019) of all nationalities now regularly come to attend 'naturotherapy' classes. The largest group amounted to 150 participants in July 2019. Some sleep on the beach, others rent rooms.

So far, there has been a balance between the small, indigenous population, offering basic tourism facilities, and low-demanding travellers. As beaches are protected from an environmental point of view and all taverns and bungalows are technically illegal buildings, the tourism offer in this respect remains uncertain, low and local. But the new municipal team elected in spring 2019 aims to legalize the island's tourist facilities on the beaches and this could offer the possibility for families to sell their businesses and land to developers. Some investors could be interested, despite the swell and the uncertainty of ferry services.

7.5.3 Gavdos and its kavatzas, heterotopias of juxtaposition

Gavdos is Greece, but this is not so self-evident on the beach. Sarakiniko, with the largest number of taverns, has the uncertain seaside allure of parts of South America or of poor Caribbean islands. The confusion is enhanced by dreadlocks and music of all kinds, from reggae to lounge. Most of Europe's nationalities live side by side on Gavdos, and, as already noted, there is great social diversity, but you cannot see it at first sight, because everyone on the beach is naked, casual and open-minded. On the beaches, everything is smooth, pleasant, easy going. But beaches and kavatzas nevertheless break up entrenched binaries such as private/public, ideal/real and above all nature/culture.

When you leave the beach and come across a very old lady, all in black among farm ruins in the heights of the island, you suddenly realize another juxtaposition: an ancient world (a rural world with strong social rules and constraints) ends up disappearing right next to the new one of the beaches, somehow postmodern. A transitional world is missing in between!

7.5.4 Gavdos and its kavatzas, heterotopias of time

No factory, no activity in the fields, no construction site. As time is often measured by activity, 'time doesn't exist'. This was often said by interviewees – about one in three. When it flows, time flows slowly. Everyone agrees that the days go by without even noticing. Most interviewees say the same thing: getting fresh water, cooking, washing clothes, all of this takes time, a lot of time in the context of Gavdos beaches. What's left is used for bathing, chatting, having fun.

Above all, as everybody feels free doing whatever they want, there is no plan for tomorrow, no appointment, no strict scheduling, no fragmentation of time. There is only one time: the present that allows people to perceive a flow of experiences, not only a flow of seconds. Varvarousis (2012) notes that the slowing down of the pace of life, the fact that there's plenty of spare time for personal and collective activities as well as the absence of compulsive planning of the daily routine, are important features that characterize the eco-community of Gavdos.

7.5.5 Gavdos and its kavatzas: open or closed heterotopias of initiation?

Bigell (2007) mentions that islands and deserts are well suited to function as heterotopias because they present physical barriers to the space of a cultural centre. The island is both 'closing' and 'opening'. Prison is indeed a permanent vocation of the island, although for the continentals, island means freedom, nudity, a privileged place of naturalism and permissiveness. Gavdos is an island that is furthermore alternatively accessible or cut off from the world, depending on the swell, and its beaches are alternatively heaven or hell, depending on the wind. These two factors limit visits to regulars: motivated or, more often, invited people. In many cases, spending time on Gavdos is a proposal or an injunction from friends: 'I was told: you absolutely have to go to Gavdos!', or 'A friend told me: Gavdos is a place for you!'

On the beaches, everyone is open-minded to those who are open-minded, but everyone reserves the right not to participate if 'not in the mood'. There are only a few rules to respect in order to be accepted:

- 'First come, first stay!'. This rule is supposed to apply to all kavatzas, even those well arranged.
- 'Don't cut a single cedar twig, even if it seems to be dead! It can be revived after years!'
- 'For the shit, a deep hole in the sand, never in the same place!'
- 'Don't pee near a kavatza!'
- 'Take home all your garbage!'

A kavatza which is at the same time open (no key, no fence) and closed (private space is delimited by stones) concentrates the oppositions exposure/hiding place and appearance/disappearance. A first stay is often an initiation with two degrees:

a. First degree: confirm that you are able to live within nature. You will have to find the right comfort, the right shelter from the sun and the wind. You will have to get water, cook in basic conditions, stay clean, get rid of a lot of habits and find pleasure in all this.
b. Second degree (optional): share community moments on the beach, in akavatza-kafenion or around groups of musicians at night.

7.5.6 Gavdos and its kavatzas: heterotopias of perfection and illusion

Bigell (2007) mentions a common perception of nature in the 18th century as an experience of edification: 'nature was a place for meditation, a source of joy and pride, a teacher, a bridge to the divine'. This is still especially true for Gavdos where the search for authenticity includes rites of passage (see above: degrees of initiation) connected to nature and leads to 'freedom' (the word most heard during the interviews) through sensitive bodies. Nudity is linked to the concept of health, extended to wholeness and holiness.

Three basic motivations can push a traveller to Gavdos – the weight of each varies, according to his/her profile: a) rejection of meaningless materialism; b) search for harmony with nature; c) participation in an anarchistic social experiment. This has to do with personal development, new-ageism and postmodernism. Fortunately, assumed pragmatism prevents people on Gavdos from losing themselves in vain efforts. It is interesting to note that the rejection of materialism does not lead to the banning of mobile phones, computers, solar panels to operate them, or tarpaulins made from sophisticated materials to shelter from the wind (discomfort is not sought in the kavatzas). In the same way, the search for harmony with nature does not lead to not trampling the roots of the cedars and only finding food by hunting, fishing or gathering. Participation in an anarchist social experiment does not require giving up the health and safety benefits of developed societies where travellers originate. And finally, idealism aside, a minimum amount of money is required to stay very long, even though earning it is recast in terms of language. Here is a young seasonal community member, a part-time craftsman in Agios Ioannis: 'In your language I have a job but, in my language, it's holiday all the time. I play. I work. I just do certain things to make money.'

7.6 Conclusion

Tourism on Gavdos gives some idea of what degrowth in tourism might be like: little infrastructure, few goods and services, frugality, spending cuts, slow pace and commitment over time. Gavdos attracts alternative tourists who reject materialism, as well as conventional tourists in search of freedom, nature and authenticity. The physical, economic and social context of the island – resolutely at the margins – allows and integrates, up to now, this attendance. Tourists and local people operating in tourism share, in a way, the same interest in not letting mass tourism emerge. The former could lose the free spaces they appreciate so much, and the latter could have to fight with competitors and lose control of island affairs. But during the interviews, there was never any mention of any associative structure or lobby group whose aim would be to promote a certain kind of development for Gavdos. The fact that tourism is almost the only activity on the island is another weakness in terms of balance and sustainability. Almost nothing is produced locally either by those who are from the island or by those who have settled permanently on the beaches, and all foodstuffs arrive directly from Crete by ferry.

Nevertheless, tourism on Gavdos is resilient. Accommodation and taverns are family-run. In the event of a crisis such as the COVID-19 pandemic, there are no heavy investments to pay off or inactive personnel to manage. Furthermore, most of the clientele is Greek and can be back as soon as the crisis ends. But a place that escapes trivialized tourism becomes attractive to tourists looking for original destinations. In recent years, framed groups (in a commercial context where nature and isolation are commodified) have got into the habit ofspending time on the island for personal development courses. Furthermore, technology allows now frantic tourists (who want to be able to say, returning home: 'I've been to the southernmost point of Europe') to spend only one day on Gavdos, either via speed-ferries that run on certain days in summer or fast boats that come from Crete, sail around the island and drop off their guests on the less frequented beaches, like Potamos. If tourists become too numerous this may be problematic. Above a certain attendance, ecological disorders will arise on the beaches, linked for instance to the permanent trampling of the woodlands or the use of nature for latrines. This risk could increase after the COVID-19 pandemic if one of the effects of this crisis is to redirect to less distant destinations and to free camping, Greek tourists usually consuming other forms of vacation. A record crowd has already been observed on Gavdos after the outbreak of the Greek economic crisis.

People who return regularly to Gavdos, often for long periods, who sometimes settle permanently there, have various profiles, but similarities emerge. A stay on Gavdos is a way to feel free from materialistic and social constraints, to connect or reconnect with nature and to learn from this unique experience. This applies to the marginals who build huts on the beaches as well as to the conventional tourists who share the same spaces. These similarities are also the core values of a community with strong bonds between individuals, which dominates social life on the beaches. Andriotis (2013) used the adjective 'antinomian' to describe this community. If this term seems to be still appropriate, interviews conducted in 2018 and 2019 for the present study reveal a less homogeneous community than it appeared in 2010 and 2011. Various reasons could explain this difference: a) methodology: the sample of Andriotis was only Greeks, interviewed in Greek during July or August while the sample in this study took into account people of all nationalities, interviewed in English or French during spring or late summer – this could have affected the findings as well as differences in the nationality and culture of the authors; b) timing: the study of Andriotis was conducted during the Greek economic crisis and this might have oriented the responses of the interviewees in the direction of radicalization against capitalism and established norms while other concerns appear in the present study: environment and personal development. In a sense, it is not surprising if the profile of those who attend Gavdos changes over time, whether it is because the world is changing or because individuals are changing.

Difficult to access, by turns open and closed, Gavdos is an island inhabited in an original way, a space where kavatzas – these tiny dwellings on the beaches, in the shade of providential, almost sacred cedar trees – are the most evanescent and most significant elements. These natural shelters that allow free camping on the beaches are the main mode of accommodation of the island, more than ten times as numerous as the rooms offered by tourism facilities. But kavatzas belong more to the dream world of those who inhabit them for a while than to the referential universe. In this sense, they are effective catalysts to generate heterotopias, i.e. realized utopias with different rules from the usual world. Thus, the principles that Foucault (2001) listed to characterize a heterotopia are verified for each kavatza as well as for each beach or for the island of Gavdos itself.

Despite this heterotopic dimension, kavatzas can become economically coveted objects as there is a real demand for this type of original tourism. This is revealed by an anecdote reported by a 40-year-old Gavdos regular, confirmed by a 30-year-old tavern manager from Sarakiniko. In early summer 2019, a young Cretan woman set up a dozen empty tents with a companion, privatizing Sarakiniko's best kavatzas. She then offered them for rent at 23 euro per night on the Airbnb platform. The regulars of Gavdos asked her to dismantle the tents and respect the spirit of freedom of the beaches of Gavdos. As she refused, the tents were dismantled by force. There was a fight and the island policeman had to intervene, which is quite exceptional. In the end, she stayed with two tents, one to sleep in and the other to offer Thai massages.

References

Andriotis, K. (2013) The 'antinomian' travel counterculture of Gavdos, an alternative mode of travelling. *Annals of Tourism Research* 40, 40–58.

Bigell, W. (2007) Cultural significations of heterotopian spaces of nature. *American Studies in Scandinavia* 39(2), 1–13. DOI: 10.22439/asca.v39i2.4604.

Crognier, E. (1994) *L'écologie humaine*. PUF Que sais-je, Paris.

Deleuze, G. (1953) Causes et raisons des îles désertes. In: Lapoujede, D. (ed.) *L'île déserte et autres textes*. Editions de Minuit, Paris.

Doyon-Gosselin, B. and Bélanger, D. (2013) Les possibilités d'une île. De l'utopie vers l'hétérotopie, in temps zero, n°6 [en ligne]. Available at: URL://tempszero.contemporain.info/document956

Foucault, M. (2001) *Des espaces autres", In Dits et écrits*. Gallimard, Paris.

Foucault, M. (2009) *Le Corps utopique, suivi de Les Heterotopies*. Nouvelles Editions Lignes, Clamecy.

Galanakis, M. (2009) *Gavdos*. Ed Galanakis (in Greek), Rethymno.

Harvey, D. (1989) *The condition of post modernity*. Blackwell.

Jacquard, A. (2004) *De l'angoisse l'espoir. Leçons d'écologie humaine*. Calmann-Lévy, Paris.

Johnson, P. (2013) The geographies of heterotopia. *Geography Compass* 7(11), 790–803. DOI: 10.1111/gec3.12079.

Lonely Planet (2020) Gavdos Island. Lonely Planet. Available at: https://lonelyplanet.com/Greece/gavdos-island (accessed 8 January 2021).

Melay, A. (2019) Au-del de l'utopie: l'île comme révélateur de l'état socio-géopolitique du monde actuel. *Carnets [Online]* Deuxième série-17 | 2019.

Morin, E. (1979) *Le Paradigme Perdu: La Nature Humaine*. Seuil, Paris.

Rieucau, J. and Lageiste, J. (2008) La plage, un territoire singulier: entre hétérotopie et anti-monde. *Géographie et Cultures* 67, 3–6.

Soja, E. (1989) *Postmodern Geographies: The Reassertion of Space in Critical Social Theory*. Verso Press, London.

Stefanakis, C. (2003) Stefanakis, C. (ed.) *Gavdos, the Emerald of the Greek South*. Thessaloniki.

Van Campenhoudt, L., Marquet, J. and Quivy, R. (2017) *Manuel de recherche en sciences sociales*. Dunod, Paris.

Varvarousis, A. (2012) Eco-communities in Greece and the project of a sustainable degrowth society. *Contribution to the 3rd International Degrowth Conference for Ecological Sustainability and Social Equity in Venice*, Degrowth.info.

8 Pushed Over the Periphery: Downsides of Degrowth on a Small Island – Experiences of Tourism Degrowth on the Isle of Man

Brendan Canavan*

University of Huddersfield, West Yorkshire, UK

8.1 Introduction

Considering the Isle of Man, a small island that has largely moved on from tourism following a long period of decline of its once significant tourism industry, this chapter looks at how small islands are exposed to issues of peripherality due to their limited size, frequent isolation and accessibility issues. In such places, tourism development can overwhelm unique and fragile local landscapes. Degrowth may be a means of better managing situations where tourism risks taking over. However, as shown in this study, tourism degrowth itself can undermine local landscapes.

Research conducted with Isle of Man tourism stakeholders (see Canavan, 2014) aimed to explore current attitudes towards the tourism industry on the island. The broad impacts of the industry on island life and landscapes were discussed. Revealed was the social and environmental significance of long-term tourism decline in a location with a long history of hosting tourists. Although the Isle of Man has moved on economically, after a difficult period of transition between tourism and offshore finance industries, the social and environmental impacts of the loss of tourism are more persistent. Consequently, concern is for the ongoing stagnation of tourism.

The downsides of tourism degrowth are explored in the wider context of small islands, whereby economic, social and cultural challenges faced by the inhabitants can be exacerbated by the loss of tourism. Small island communities, due to their peripherality, face heightened challenges to habitation.

*b.canavan@hud.ac.uk

DOI: 10.1079/9781789245073.0008

A lack of alternative development options moreover means they may be particularly dependent upon tourism. In such places, decline may thus call into question the continued existence of communities.

The significance of this study is twofold. First, attention is drawn to the experiences of small island stakeholders and inhabitants. These are often overlooked in tourism research. Second, notes of caution regarding tourism degrowth are sounded that challenge or at least add nuance to arguments advocating this. A call for sustainable management of degrowth is made if small islands are to avoid being pushed over the periphery.

8.2 Small Island Tourism

Small islands have a satisfying self-containment. Their distinctiveness is bounded by a maritime boundary. Contained within the shorelines of small islands are uniquely adapted natural and cultural ecosystems. Islands are renowned for their natural endemism; the prevalence of unique species that occur on islands thanks to evolutionary isolation (Berry, 2009). They also feature significant cultural endemism. Cut off from adjacent mainland areas, island ways of life evolve in relative isolation. This may be seen in the persistence of remnants of cultures lost elsewhere, such as the Gaelic languages still spoken on islands of Brittany, Scotland and Ireland (Ball and Muller, 2009).

In addition, peripherality sees unique cultural adaptations in response to the pressures and possibilities of island life. Living with peripherality nurtures adaptability and hardiness. Small islands often feature high levels of entrepreneurship from inhabitants used to exploiting limited opportunities (Baldacchino, 2015). Chaperon and Bramwell (2013) describe for example the reputation of Gozitan islanders for economic resourcefulness and creativity, such as tax avoidance, in the face of considerable economic constraints. Vibrant art, literary and music scenes can moreover be found where inhabitants need to entertain themselves. Belying their small populations and geographic isolation therefore, islands are often unexpectedly vibrant and creative, as well as distinctive, places.

It is this peripheral other-worldliness, and the unique responses this engenders from nature and people, which makes islands so attractive to tourists. Isolation protects island landscapes from intrusion and creates conditions for unique sociocultural development. 'Ironically, the very consequences of lack of development, the "unspoilt" character of the landscape and distinctive local cultures, become positive resources as far as tourism is concerned' (Duffield and Long, 1981, p. 409). Small islands appeal to visitors because of their rich natural and cultural resources. Globally, islands punch well above their little landmasses in tourism terms. In many respects small islands are overlooked. However, in relation to tourism they have variety and longevity of experience hosting guests, and subsequently significant insights to share with the industry.

Tourism on small islands is nonetheless unpredictable, as elsewhere. Indeed this may be more so given their relatively limited resources, autonomy, recognition and hence ability to take charge of their own destiny. Various previously successful destinations, such as the Isle of Man, have fallen from favour, and consequently experienced degrowth, whether intentional or not. The ability to manage or adjust to such fluctuations may similarly be restricted by the inherent limitations of small islands. The urgency of anticipating and understanding degrowth in small islands as both likely and of heightened impact is therefore clear.

8.3 Pushing Back the Periphery

Tourism can be especially important in small islands as the industry helps to combat the difficulties of life in a peripheral area. Peripheral locations are termed such because they are disadvantaged in being distant from population and economic centres, and travel to them can be time-consuming and costly (Nash and Martin, 2003). Socially for example, isolation can result in a lack of relationship and entertainment opportunities (Marjavaara, 2007). Culturally, islanders can feel cut off and excluded from wider debates (Canavan, 2013). Economically, limited development options are due to restricted economies of scale and transport costs, and small territory that can be put to few economic uses (Aramberri and Butler, 2005). In addition, limited human, financial, natural and infrastructural resources lead to import dependencies (Ayres, 2000). Thus, small islands suffer high leakage rates, explains Aramberri, whereby money flows out of locations in order to pay for products and services not available locally.

Andriotis (2004, p. 115) summarizes these limiting consequences of peripherality: 'By their very nature islands face a number of inherent disadvantages. They are small in size with declining populations; they suffer from isolation, peripherality, external dependency and diseconomies of scale; they are rural in character; and they have a scarcity of resources, meaning mainly that their alternatives for industrialisation and self-sustaining growth are limited.' It is clear that due to their relative geographic isolation that cuts them off from wider systems, as well as their smaller size which limits availability of local resources, small islands are unique and special. Yet they simultaneously face significant economic and social challenges.

Accordingly, the viability of small island communities can be uncertain thanks to the constraints imposed by peripherality. As Keane *et al.* (1992, p. 407) explain: 'The problem of sustaining population on small coastal islands is a difficult one. The difficulties are created by the limited resource base and consequently limited employment opportunities, by the restrictions which size can place on the provision of services, and by the problems of access to markets and to the mainland.' Indeed, many small islands worldwide have witnessed population ageing and decline, if not total abandonment, as residents are attracted by greater educational, financial and social opportunities abroad. The scale of out-migration in the British Isles for example, is highlighted by Berry (2009). He details that during the 19th and 20th centuries, four of the Orkneys have been abandoned, 12 of the Inner and 15 of the Outer Hebrides, and 145 of the islands of Ireland.

In the face of this habitation challenge, tourism as a route to social and economic progress has been pursued by many small islands worldwide. Tourism can be a relatively easy industry to establish on small islands as it requires only basic human and infrastructural resources, suiting an island's limitations in these respects. Rather, tourism is dependent on the quality of natural and cultural resources that small islands have in abundance. Hall and Boyd (2005) suggest that the restricted number of development opportunities in peripheral zones makes tourism a critical economic tool. Tourism is moreover an industry that can actively address peripherality in that the sector brings a wide range of interrelated benefits. Puhakka *et al.* (2009, p. 531) summarize the positive aspects of tourism development in peripheral areas as 'providing an alternative to fading traditional industries, stimulating infrastructure development, fostering pride pertaining to culture and community, promoting cultural understanding among visitors, preserving cultural heritage, promoting cross-institutional understanding and creating educational opportunities.'

Consequently, tourism may be a valuable source of inputs that can help to sustain small island populations and address pressures of outward migration that threaten to erode the continuation of their unique communities. Ireland (2003) observes for instance, how tourist interest in lifestyles of indigenous peoples has helped to give such communities an important source of income and integration with wider regional communities and financial systems, with growing self-respect and political influence resulting from this. Economic, social and cultural opportunities can stimulate, revive and preserve island life. As such, Péron (2004, p. 336) concludes that on the islands surrounding the Breton peninsula tourism 'has created a strong and original identity for them that is some consolation for their decline in numbers and makes the otherwise hard life on an off-shore island bearable'.

8.4 Why Degrowth?

In light of these highly relevant contributions, expansion rather than contraction of tourism on small islands would appear to be imperative. Indeed, this is frequently the case and tourism has been pursued as a route to development by many small islands worldwide. As is widely appreciated however, alongside diverse benefits, many negative impacts can be caused or worsened by tourism development. In spatially confined locations, resource demands and development impacts can both be heightened. Various cases have illustrated, for example, the potential for tourism to overwhelm, displace or commodify cultural resources. Thus, Royle (2003, p. 27) argues that the tourist positioning of Irish islands has driven 'a process that has rendered the island into little more than a living, interactive museum of itself with islanders as actors'. Indeed, overtourism can threaten to erode the distinctiveness of islands, which in itself makes them special places to live and to visit. What is more, the reliability of the tourism industry on small islands is questioned. Tourism is often highly seasonal, and it can be difficult to cope with either the volume of visitors during peak times, or the lack of demand during down periods. Employment, infrastructure use and business support may be unpredictable and only for certain periods of the year. Baum and Hagen (1999) discuss the limitations of seasonality in peripheral areas and the mixed success of attempts to address this through extending the tourist period.

The famous case of the tiny Pacific island republic of Nauru in Micronesia serves as a reminder that islands are vulnerable to short-term overexploitation, by any industry. Also, of the transience of many such industries, which are able to move on after destroying long-term economic, environmental and social sustainability, whereas inhabitants themselves are not. In the 1970s Nauru experienced an economic boom based on phosphate mining. This industry exploited guano reserves, built up on the island over millennia by the migratory seabirds that nest on the isolated outcrop. Nazzal, 2005 describes the ephemerality of the industry that left devastation in its wake. The brief economic prosperity brought by mining, with islanders at the time some of the richest in the world per head, rapidly declined as phosphate reserves became exhausted.

In Nauru an environmental and socioeconomic disaster has been left behind. This is described by Fagence (1997), who outlines how mining has left the island's interior a devastated, uninhabitable moonscape. A narrow coastal band, the only area untouched by mining, now hosts the increasingly overcrowded population. Loss of arable land means residents are dependent on imports, particularly of fatty and sugary canned goods, which has in turn contributed to increasing obesity. Loss of natural forest cover meanwhile has led to a disastrous and total disruption of island ecosystems, reduced rainfall, humidity, and increased occurrences of drought. For Nazzal (2005: 6), 'it is undeniably clear that alongside the forceful erosion of the land, the Nauruan way of life and intricate relationships with their surroundings was also eroded'.

Consequently, many case studies call attention to the need for careful management of development in small islands and peripheral areas, particularly fragile as these are. Highlighting the need to protect environmental resources, Jaafar and Maideen (2012) suggest for example that demand should not outstrip supply in Malaysian small islands. Tourism is often characterized as bringing short-term economic stimulus at a significant environmental cost (Sharpley, 2003). Tapper (2001) believes that these issues make tourism a short-term industry. Meanwhile Kontogeorgopoulos (2005) contends that tourism should be regarded as a transition sector in the sense of this being a route to development for less developed locations.

If notes of caution concerning tourism development are well established, then degrowth has more recently emerged as an area of discussion in considering how to manage what can be an unsustainable sector. A number of authors, such as Kallis (2011), have suggested degrowth as an alternative to complement more traditional sustainability perspectives. Within tourism studies, advocates suggest that fragile destinations might seek to reduce tourism, in order to lessen the harmful impacts of hosting visitors. In certain circumstances this may be advisable. Yet undertourism leaves islands exposed to the threat of socioeconomic stagnation and outward migration. Tourists bring a lot of positive inputs which can contribute to island destinations as more vibrant, diverse and enjoyable places to live, work and play. Degrowth can therefore pose a threat to the long-term viability of island communities and push these over the periphery. Considering one such island in more detail helps to illustrate how degrowth needs careful consideration in peripheral areas.

8.5 Isle of Man Tourism

The Isle of Man is located midway between Britain and Ireland in the centre of the Irish Sea. Often contested by its larger maritime neighbours, the island maintained its status as a largely independent Norse-Gaelic kingdom until 1266. A high degree of autonomy has remained since, with the island self-governed by the Viking-established Tynwald Parliament, that is locally reputed to be the oldest in continuous existence. Around 80,000 people live on the island, with around one third of these based in the seaside capital Douglas (Isle of Man Census, 2016). Typical of many small islands, the peripherality of the Isle of Man, distant from population centres and between competing cultural-political powers of the British Isles, has resulted in an idiosyncratic Manx culture, manifest in a variety of cultural sites and events. These include the TT festival. Over a century old and the world's longest running road race event, this is the highlight of the tourist calendar. The island also has a diversity of natural habitats: coastal plain, dune, heath and cliffs, with wooded glens cutting into central moorland that rises to just over 2000 ft (600 m) at the highest point, Snaefell. These support diverse and at times uncommon resident and migratory flora and fauna, including important populations of rare birds, such as the hen harrier and chough.

By the late 19th century, the Isle of Man had established a reputation as a somewhat exotic and upmarket tourism destination. The requirement of a sea crossing to reach the island was turned to marketing advantage; positioning as a more exciting location than mainland resorts. The island's cultural and natural assets were made the most of with a wave of tourism infrastructure-building. This included steam and electric railways to connect the island's different towns, each of which could offer unique attractions, activities and atmosphere. Douglas Bay, with an impressive promenade of multi-storey guest houses and shops. Peel and Castletown having historical centres dominated by ancient castles. Ramsey fronting onto a large sandy bay. Port Erin featuring a scenic beach setting framed by hills. Elsewhere, pleasure gardens were set up in various glens and the island's moors became popular with walkers. A variety of natural and cultural scenery fills an island little more than 15 miles wide and 30 miles long.

From the 1890s to the 1960s, the Isle of Man was a significant tourism destination, typical of the seaside and wellness resorts of the period. However, from the mid-1970s onwards, the industry experienced a rapid downturn. As with other traditional resorts of northern Europe and beyond, the struggle was to compete with more exotic, fashionable and climatically favourable overseas locations opened up by advances in air travel (Walton, 2000). The failure of the Isle of Man to adapt to and counter this trend has been documented by researchers such as Cooper (1995) and Baum (1998), who highlight the lack of investment in infrastructure and innovation, failure to diversify markets and an overreliance on a traditional, increasingly dated image. The ability of a small island with few resources to counter large-scale global changes in tourism tastes and economics is however limited regardless of historic missteps, such as the failure to adequately preserve underutilized buildings and facilities of historic interest and cultural significance.

In place of tourism, offshore industries now make up the majority of the Manx economy. This umbrella term encompasses the financial sector. The island is, like many other peripheral nations, a tax haven, offering financial services to the international wealthy and global business. Brown (2010) even suggests that peripheral locations have few alternatives to tourism other than finance, where permitted by local legislatures, and then smuggling-type activities where not. Attracted by low tax rates many organizations have subsidiaries in the island, as a patrol of the small plaques beside the front doors of buildings in the capital's business district attests. Alongside financial services are sectors such as shipping and aircraft registration, gambling and other activities that appreciate hands-off local regulation.

In many respects, the Isle of Man's move beyond tourism has been a successful one. The dominant financial services industry has brought significant wealth to the island. Local incomes, GDP per head, economic growth and unemployment rates compare favourably with the wider British Isles (Isle of Man in numbers, 2019). This is in contrast to the situation in the 1980s, when tourism had entered terminal decline and the financial services industry had not yet boomed. Then, GDP on the island, as a percentage of GDP per head in the UK, declined from 88% in 1970, when the tourist industry was still significant, to below 60% in the early 1980s when it had greatly declined (Rawcliffe, 2009). At this decade's nadir, the island's population had fallen by around a quarter as people were forced to emigrate in search of work. Older residents remember this difficult period. They describe collapsed property prices meaning people trapped in homes they couldn't sell, the loss of community facilities such as cinemas and dance halls, and family members forced to seek work elsewhere. However, the rebalancing away from tourism has had a number of cautionary impacts.

8.6 Downsides of Degrowth

From the Isle of Man perspective, the immediate consequences of rapid, prolonged and significant tourism degrowth were highly negative. This is the situation for small islands more broadly, where socioeconomic problems of tourism degrowth include falling employment and reduced economic opportunities. Correspondingly, indicators such as high unemployment levels and benefit claimant rates are of social deprivation. Getz (1994) talks more explicitly about tourism decline, illustrating how the closure of facilities has disillusioned residents in a rural Scottish community and made the area a less attractive place to live. Similarly, a sense of disappointment and pessimism in stagnant Baltic Sea destinations is highlighted by Twining-Ward and Baum (1998). Meanwhile, reduced cultural, entertainment and social facilities for young people in Baltic islands is linked to loss of tourism by Baldacchino and Pleijel (2010) and Marjavaara (2007).

Having gone through such a period of difficulty, Isle of Man politicians' and residents' gratitude for the offshore finance industry improving this situation is unsurprising. Over the past three decades, inward migration, house prices, employment and GDP have generally and consistently improved as offshore finance has grown. However, this appreciation is at times to the extent that it can be difficult to criticize the sector. It downplays the dependence of the island on a controversial industry. Shaxson (2011) reviews in detail how tax evasion and avoidance enabled by offshore financial centres has a significant impact on government revenues, particularly in developing nations.

Regarding overdependence, the term 'political capture' is coined by Hampton and Christensen (2007), referring to how the powerful offshore finance sector may be an issue on small islands where policy comes to be co-opted by the industry. They highlight the potential for democratic processes (which often have dominant or no political party representation, a narrow pool of aspiring politicians, close personal networks and limited free press) to be subverted by powerful interest groups. Where offshore finance becomes ever more dominant it can be difficult to balance the interests of the industry with wider issues such as conservation. Dependence on one industry moreover leaves islands exposed to a potential downturn. Richardson and Fluker (2004) characterize island economies as vulnerable to crisis due to their limited diversification and influence over wider regional and global trends. The European Union, OECD and other transnational organizations have pushed for increasing regulation of tax havens and it may be that their future is uncertain in the face of such pressure.

Also significant on the Isle of Man is the loss of positive impacts associated with tourism. Economically, the financial services sector provides plentiful and frequently well-paid jobs. However, wider impacts of tourism have been less readily compensated. Socially there has been a decline in tourism-related infrastructure and facilities often used by communities also (Canavan, 2013). This combines with lessened social opportunities as fewer tourists mean fewer chances to interact with outsiders. Additionally, there is a decline in the overall atmosphere of urban centres traditionally animated during the summer by visitors. Environmentally meanwhile, concern is for the loss of natural sites with demands from the financial sector for housing and commercial developments. Where tourist interest provides an argument for valuing natural scenery, then so too its loss undermines advocates. Finally, the experience of the Isle of Man highlights how negative cultural impacts of tourism degrowth can be significant. Here the decline in tourism appears to have affected islanders' self-image. Reduced outsider interest may lead to a growing sense of isolation. Whereas tourist interest confers status upon a location, tourism decline suggests that a place is too peripheral to be visited.

Together, these issues potentially make the island less appealing to residents and visitors alike. To illustrate, the loss of tourism infrastructure, including attractive and historic buildings associated with tourism, has arguably damaged the diversity and identity of the island (Canavan, 2014). As Hall and Zeppel (1990) describe, architecture is closely linked to community identity, and the loss of local vernacular can damage sense of place. Péron (2004) describes the same regarding natural scenery in small islands, highlighting this as vulnerable to economic upheaval, decline or changes to new economic practices, because these may make obsolete or conflict with traditional land use patterns. Van den Bergh (2011) cautions that individual and collective responses to degrowth are complex, and such a process may therefore have significant social as well as economic costs. This appears to be at least somewhat the case on the Isle of Man.

8.7 Degrowth Disruption

It is the disruption of degrowth which underpins its potential downsides. Losing tourism has implications for communities which have evolved alongside, adapted to and incorporated the sector into their lifestyles. It has been suggested that destinations with long histories of hosting visitors may develop a 'tourism culture' whereby tourism becomes a part of everyday life. Such tourism culture can be seen as a nexus between hosts and guests that evolves over time into something unique and distinctive (Canavan, 2016). This combines host culture – the cultural practices indigenous to a locale, its particular arts and crafts, language, stories, festivals and ways of doing things – with tourist culture, which refers to the background of guests, who, informed by their originating cultures, bring their own ways of being and modes of expression with them on holiday. It is the interplay between these two which can lead to an emergent tourism culture that is informed by both.

To illustrate, host cultures may evolve alongside, adapt to the presence of and co-opt guest cultures. Tourists are, for example, typically less restrained and more hedonistic when on holiday, and bring with them associated attitudes and behaviours. These can rub off on a locale which may over time become more liberal as a result. In addition, the nature of tourist hosting, involving interactions and exchanges, means destinations may be particularly multicultural, open-minded and dynamic. New migrants attracted by tourism bring new inputs into the local economic and social landscape, for example. Similarly, long-term residents may develop entrepreneurial skills as they pursue opportunities within tourism (see Chaperon and Bramwell, 2013). Indeed, the co-creation of hosts and guests in terms of developing tourism products and experiences, adding atmosphere, valorizing cultural assets and revitalizing existing products is discussed by Richards (2014), while Crespi-Vallbona and Richards (2007) highlight how such co-creativity can reinvigorate local culture for both residents and tourists.

The consequence of such host–guest interactions is therefore cultural evolution and stimulation. For example, cycles of seasonality linked with tourism may alter local sociocultural landscapes, leading to adjusted family routines or traditional calendars as individuals try to exploit the peak season. Thus, tourism potentially becomes over time an integral part of local landscapes and identity. This is demonstrated by Zhihong (2007) in relation to the Bai Chinese indigenous community, where tourism has become part of their culture and part of their ethnic identity. Social roles, events and routines all evolve because of and alongside tourism development. So too do local facilities, infrastructure, natural and cultural landscapes and even the sense of local identity, all moulded by the mutual exchanges between hosts and guests over time.

The presence of tourists causes a society to adjust its routines in order to accommodate them and to analyse itself through appreciation of what incomers report back. This is important for cultural renewal. To illustrate, visitors' interests in natural and cultural resources may stimulate fresh appreciation of these. The sustainable tourism literature credits an expansion of the conservation ethos linked to hosting guests. Communities may, for instance, be encouraged to re-evaluate the use of natural resources in light of the financial and social inputs brought by tourism interest and example (i.e. Reimer and Walter, 2013). Tourist presence is also an opportunity to share local culture. Communities can be motivated by outsider interest to explore, revive and reinterpret traditional aspects of local identity. An example of this complex dialogue is Maltese handicrafts, with some products and styles originally intended for external audiences being appropriated into local culture (Markwick, 2001).

Hence tourism is valuable for both guests and hosts as it can trigger reflexivity, foster connections and introduce new ideas, that together expand horizons. Tourism cultures are valuable. If they are disrupted by degrowth, then they can be undermined and eventually lost. With a decline in visitor arrivals the linkages between peripheral areas and wider networks can be lost. Sociocultural exchanges diminish. Sense of isolation and neglect can increase. Therefore, it is not only economic disruption, but social, natural and cultural commotion that can accompany tourism degrowth.

8.8 Conclusion

The implications of research on the Isle of Man are that tourism degrowth can herald diverse and enduring impacts. Although tourism has long since declined on the Isle of Man, inhabitants remain uncomfortable with its loss and would like to see a return to a more touristic state. Going beyond mere nostalgia, various negative social and environmental consequences of tourism degrowth are perceived. These range from the specific, such as the loss of particular facilities, to the more esoteric, such as a sense of greater disconnection from overseas cultures. What makes this critical perception of tourism degrowth particularly interesting meanwhile, is that tourism has been successfully replaced over the previous four decades by a highly successful economic model that has brought financial prosperity to the Isle of Man. Highlighted therefore, is that even where the initial disruption of degrowth has been adjusted to, longer-term issues may persist. Tourism is perhaps an industry which can be appreciated more deeply by community members than is sometimes given credit. Likewise, its loss can be mourned accordingly.

The experience of the Isle of Man highlights how tourism degrowth can have extensive negative impacts on a locale. This is across interrelated economic, social and environmental domains. Problems of industry decline may be especially marked in peripheral locations such as small islands. The habitability of these places is never easy and may be further undermined by the loss of the inputs that tourism brings. Furthermore, the negative impacts of degrowth may be felt even where a location is able to move on to other economic sectors. It is often the case that tourism becomes an integral part of local day-to-day life and, beyond that, culture. Residents interact with tourists, use tourist facilities, benefit from shared infrastructure and participate in the industry in a variety of ways. Tourism degrowth and transition to alternative industry can consequently disrupt the impacts, exchanges and mutually evolved landscapes of destinations, as indeed has been the Isle of Man experience. Findings from the island highlight that the decline phase of tourism poses its own set of challenges to local communities. Tourism as an industry brings many relevant benefits to the Isle of Man, and its downturn has seen a reduction in these. Degrowth engenders economic, social, and environmental upheaval. Traditional employment, social and familial roles may alter. Facilities, architecture and landscape use change. And internal and external identities of place shift. Such rapid change can be traumatic for a community, and sustainability of cultural and natural environments may be threatened in such a context.

The Isle of Man thus serves as a warning of the negative consequences which may accompany tourism degrowth. Research on the island highlights that tourism decline can negatively impact on a variety of economic, social, cultural and environmental criteria. In turn, this may undermine local habitability. The quality of urban and rural landscapes has been harmed by tourism decline. This includes the lessening diversity of social and entertainment facilities, and loss of places associated with community traditions and memory. Moreover, fewer visitors means less atmosphere and opportunity for interactions that can be personally pleasurable and culturally affirming. Islanders can feel increasingly cut off from and discredited by wider areas. Whereas the tourism economy once integrated in broader socioeconomic systems and gave the Isle of Man a sense of status, offshore finance is less well known, often has an image problem and tends towards the needs of a transient global elite rather than fostering connections at the community level.

Although offshore finance has provided economic benefits that have rescued the island from a seemingly terminal decline linked to its tourism sector drop-off, its environmental demands and facilities support are quite different to those of tourism, creating dissonance. To illustrate some of the challenges of such upheaval, Brown (2010) suggests that many residents may be left out of the finance industry. Meanwhile Shaxson (2011) argues that residents of what he calls 'treasure islands' are likely to be displaced by new infrastructure serving international elites with little local sensitivity or nuance. For the Isle of Man, hotels previously used by local communities to host events are converted to more profitable residential or commercial use. Overall, the island's experience of tourism degrowth, involuntary and significant, draws attention to the disruption this may bring. Highlighted are the unique, relevant and positive impacts of tourism on a small island. Tourism brings more than just economic benefits, with many environmental, cultural and social ones also. These may be especially valued in peripheral locations with often restricted facilities, fewer social opportunities and prone to cultural isolation. Tourism decline can push islands further towards the periphery.

The concept of degrowth has been described as a project of radical socio-ecological transformation concentrated on the sustainable reduction of consumption and production arrangements (Büscher and Fletcher, 2017). Radical transformation may however be difficult in fragile locations such as small islands, where community resilience is already lessened. Moreover, where tourism has often become an integral part of such ecosystems over time, then industry degrowth can have wide ranging negative consequences. More subtle transition may therefore be called for when discussing degrowth. In this respect Hall (2009) explains that degrowth is a concept which is not so much connected to downsizing per se, but to the notion of rightsizing an economy in environmental terms. Meanwhile for Whitehead (2013) there are two basic antecedents on which the concept of degrowth hinges: one is that economic activity must not exceed the carrying capacity of the biosphere, while the other is that economic activity should concentrate on developing human wellbeing and happiness rather than prioritizing greed and wealth. From this perspective, degrowth refers to a transition process in which becoming or performing a steady-state economy is seen as a primary aim to be achieved (Dietz and O'Neill, 2013).

Parallels between sustainable tourism assumptions during the growth and degrowth phases of tourism industry presence are therefore clear. Sustainable principles have been applied extensively to destinations at the growth stage of the lifecycle, with the aim of protecting natural and cultural landscapes from over-development and moderating rates of change. The same is applicable to resorts in decline, where the aim should be managing over-decline and the speed of this, so that landscapes are not damaged as a result, and degrowth progresses in line with acceptable and adaptable parameters of change. Consequently, as explained by Canavan (2014), sustainable principles of stakeholder collaboration, planning, impact measurement and long-term inclusive stewardship remain relevant when reducing as well as expanding tourism. This is in order to preserve and maintain landscape integrity, diversity and attractiveness to both residents and tourists. Even more fundamentally, in peripheral areas it is to maintain viability of habitation and avoid local populations being pushed over the edge.

References

Andriotis, K. (2004) Problems of island tourism development: the Greek insular regions. In: Bramwell, B. (ed.) *Coastal mass tourism: Diversification and sustainable development in Southern Europe*. Channel View Publications, Clevedon, UK.

Aramberri, J. and Butler, R. (2005) How global is tourism? In: Aramberri, J. and Butler, R. (eds) *Tourism Development Issues for a Vulnerable Industry*. Sage, London, pp. 127–152.

Ayres, R. (2000) Tourism as a passport to development in small states: reflections on Cyprus. *International Journal of Social Economics* 27(2), 114–133. DOI: 10.1108/03068290010308992.

Baldacchino, G. and Pleijel, C. (2010) European Islands, development and the cohesion policy: a case study of Kökar, Åland islands. *Island Studies Journal* 5(1).

Baldacchino, G. (ed.) (2015) *Entrepreneurship in small island states and territories*. Routledge.

Ball, M. and Muller, N. (2009) *The Celtic Languages, 2nd Edition*. Routledge, Abingdon, UK.

Baum, T. (1998) Taking the exit route: extending the tourism area life cycle model. *Current Issues in Tourism* 1(2), 167–175. DOI: 10.1080/13683509808667837.

Baum, T. and Hagen, L. (1999) Responses to seasonality: the experiences of peripheral destinations. *International Journal of Tourism Research* 1(5), 299–312. DOI: 10.1002/(SICI)1522-1970(199909/10)1:5<299::AID-JTR198>3.0.CO;2-L.

Berry, R.J. (2009) *Islands*. Collins, London.

Brown, F. (2010) *Tourism reassessed: Blight or blessing*. Routledge.

Büscher, B. and Fletcher, R. (2017) Destructive creation: capital accumulation and the structural violence of tourism. *Journal of Sustainable Tourism* 25(5), 651–667. DOI: 10.1080/09669582.2016.1159214.

Canavan, B. (2013) Send more tourists! Stakeholder perceptions of a tourism industry in late stage decline: the case of the Isle of Man. *International Journal of Tourism Research* 15(2), 105–121. DOI: 10.1002/jtr.883.

Canavan, B. (2014) Sustainable tourism: development, decline and de-growth. management issues from the Isle of Man. *Journal of Sustainable Tourism* 22(1), 127–147. DOI: 10.1080/09669582.2013.819876.

Canavan, B. (2016) Tourism culture: nexus, characteristics, context and sustainability. *Tourism Management* 53, 229–243. DOI: 10.1016/j.tourman.2015.10.002.

Chaperon, S. and Bramwell, B. (2013) Dependency and agency in peripheral tourism development. *Annals of Tourism Research* 40, 132–154. DOI: 10.1016/j.annals.2012.08.003.

Cooper, C. (1995) Strategic planning for sustainable tourism: the case of the offshore islands of the UK. *Journal of Sustainable Tourism* 3(4), 191–209. DOI: 10.1080/09669589509510726.

Crespi-Vallbona, M. and Richards, G. (2007) The meaning of cultural festivals: Stakeholder perspectives in Catalunya. *International Journal of Cultural Policy* 13(1), 103–122.

Dietz, R. and O'Neill, D. (2013) *Enough is enough: Building a sustainable economy in a world of finite resources*. Routledge.

Duffield, B.S. and Long, J. (1981) Tourism in the highlands and islands of Scotland rewards and conflicts. *Annals of Tourism Research* 8(3), 403–431. DOI: 10.1016/0160-7383(81)90006-2.

Fagence, M. (1997) An uncertain future for tourism in microstates; the case of Nauru. *Tourism Management* 18(6), 385–392. DOI: 10.1016/S0261-5177(97)00036-8.

Getz, D. (1994) Residents' attitudes towards tourism: a longitudinal study in Spey Valley, Scotland. *Tourism Management* 15(4), 247–258.

Hall, C.M. (2009) Degrowing tourism: Décroissance, sustainable consumption and steady-state tourism. *Anatolia* 20(1), 46–61. DOI: 10.1080/13032917.2009.10518894.

Hall, C.M. and Zeppel, H. (1990) History, architecture, environment: cultural heritage and tourism. *Journal of Travel Research* 29(2), 54–55. DOI: 10.1177/004728759002900212.

Hall, C.M. and Boyd, S.W. (eds) (2005) *Nature-based tourism in peripheral areas: Development or disaster? (Vol. 21).* Channel View Publications, Clevedon.

Hampton, M.P. and Christensen, J. (2007) Competing industries in islands a new tourism approach. *Annals of Tourism Research* 34(4), 998–1020. DOI: 10.1016/j.annals.2007.05.011.

Ireland, M. (2003) Sustaining indigenous peoples in the wilderness areas of Scandinavia and north-west Russia. *Scandinavian Journal of Hospitality and Tourism* 3(1), 71–81. DOI: 10.1080/15022250310001918.

Isle of Man Census (2016) Available at: https://www.gov.im/media/1355783/2016-isle-of-man-census-presentation.pdf

Isle of Man in numbers (2019) Available at: https://www.gov.im/media/1366013/isle-of-man-in-numbers-2019.pdf

Jaafar, M. and Maideen, S.A. (2012) Ecotourism-related products and activities, and the economic sustainability of small and medium island chalets. *Tourism Management* 33(3), 683–691. DOI: 10.1016/j.tourman.2011.07.011.

Kallis, G. (2011) In defence of degrowth. *Ecological Economics* 70(5), 873–880. DOI: 10.1016/j.ecolecon.2010.12.007.

Keane, M.J., Brophy, P. and Cuddy, M.P. (1992) Strategic management of island tourism – the Aran islands. *Tourism Management* 13(4), 406–414. DOI: 10.1016/0261-5177(92)90008-U.

Kontogeorgopoulos, N. (2005) Community-based ecotourism in Phuket and AO Phangnga, Thailand: partial victories and bittersweet remedies. *Journal of Sustainable Tourism* 13(1), 4–23. DOI: 10.1080/17501220508668470.

Marjavaara, R. (2007) The displacement myth: second home tourism in the Stockholm Archipelago. *Tourism Geographies* 9(3), 296–317. DOI: 10.1080/14616680701422010.

Markwick, M.C. (2001) Tourism and the development of handicraft production in the Maltese Islands. *Tourism Geographies* 3(1), 29–51. DOI: 10.1080/14616680010008694.

Nash, R. and Martin, A. (2003) Tourism in peripheral areas? The challenges for northeast Scotland. *International Journal of Tourism Research* 5(3), 161–181. DOI: 10.1002/jtr.426.

Nazzal, M. (2005) Nauru: an environment destroyed and international law. Law and Development. Available at: http://lawanddevelopment.org/docs/nauru.pdf (accessed 6 January 2021).

Péron, F. (2004) The contemporary lure of the island. *Tijdschrift voor Economische en Sociale Geografie* 95(3), 326–339. DOI: 10.1111/j.1467-9663.2004.00311.x.

Puhakka, R., Sarkki, S., Cottrell, S.P. and Siikamäki, P. (2009) Local discourses and international initiatives: sociocultural sustainability of tourism in Oulanka National Park, Finland. *Journal of Sustainable Tourism* 17(5), 529–549. DOI: 10.1080/09669580802713457.

Rawcliffe, J. (2009) *No Man is an Island.* Manx National Heritage, Douglas, Isle of Man.

Reimer, J.K. (Kila) and Walter, P. (2013) How do you know it when you see it? Community-based ecotourism in the cardamom mountains of southwestern Cambodia. *Tourism Management* 34, 122–132. DOI: 10.1016/j.tourman.2012.04.002.

Richards, G. (2014) Creativity and tourism in the City. *Current Issues in Tourism* 17(2), 119–144. DOI: 10.1080/13683500.2013.783794.

Richardson, J.I. and Fluker, M. (2004) *Understanding and managing tourism.* Pearson Education Australia.

Royle, S.A. (2003) Exploitation and celebration of the heritage of the Irish islands. *Irish Geography* 36(1), 23–31. DOI: 10.1080/00750770309555809.

Sharpley, R. (2003) Tourism, modernisation and development on the island of Cyprus: challenges and policy responses. *Journal of Sustainable Tourism* 11(2-3), 246–265. DOI: 10.1080/09669580308667205.
Shaxson, N. (2011) *Treasure islands: Tax havens and the men who stole the world*. Random House.
Tapper, R. (2001) Tourism and socio-economic development: UK tour operators' business approaches in the context of the new international agenda. *International Journal of Tourism Research* 3(5), 351–366. DOI: 10.1002/jtr.348.
Twining-Ward, L. and Baum, T. (1998) Dilemmas facing mature island destinations: cases from the Baltic. *Progress in Tourism and Hospitality Research* 4(2), 131–140. DOI: 10.1002/(SICI)1099-1603(199806)4:2<131::AID-PTH106>3.0.CO;2-E.
Van den Bergh, J.C.J.M. (2011) Environment versus growth – A criticism of "degrowth" and a plea for "a-growth". *Ecological Economics* 70(5), 881–890. DOI: 10.1016/j.ecolecon.2010.09.035.
Walton, J.K. (2000) *The British seaside: Holidays and resorts in the twentieth century*. Manchester University Press.
Whitehead, M. (2013) Degrowth or regrowth? *Environmental Values* 22(2), 141–146.
Zhihong, B. (2007) Ethnic identities under the tourist gaze. *Asian Ethnicity* 8(3), 245–259. DOI: 10.1080/14631360701595007.

9 Freedom of Movement and Degrowth

Rasa Pranskūnienė* and Dalia Perkumienė

Vytautas Magnus University, Lithuania

9.1 Introduction

The theoretical and political fundaments that nourish degrowth can be traced back at least to the 1960s and 1970s, as Fletcher *et al.* (2019) notice, in the early voices in political and social ecology (e.g. Andre Gorz, Jacques Ellull, Ivan Illitch, Cornelius Castoriadis, Murray Bookchin and Manuel Sacristan) as well as ecological economics (e.g. Nicolai Georgescu-Roegen, Joan Martınez-Alier, Herman Daly and Jose Manuel Naredo). Degrowth was explicitly conceptualized by a network of thinkers, initially centred in France, as Paulson (2017) notices, among them philosopher André Gorz (1980, p. 13), who wrote: 'even at zero growth, the continued consumption of scarce resources will inevitably result in exhausting them completely, the point is not to refrain from consuming more and more, but to consume less and less – there is no other way of conserving the available reserves for future generations'. According to Andriotis (2014), the concept of degrowth has been introduced only recently in tourism studies. Brossmann and Islar (2019) have pointed out that while the literature on degrowth has grown substantially, there are questions about how degrowth transformations can be achieved, as degrowth calls for a society with a smaller 'metabolism' – i.e. less production and consumption to reduce the overall throughput of energy and material in accordance with ecological limits. The literature makes a strong critique of economic growth, which is seen as highly problematic and causally related to both social and ecological crises.

On 11 March 2020, the World Health Organization (WHO) declared that an outbreak of the viral disease COVID-19 – first identified in December 2019 in Wuhan, China – had reached the level of a global pandemic. The WHO director-general noted that 'we have never before seen a pandemic sparked by a coronavirus, we have called every day for countries to take

*Corresponding author: rasa.pranskuniene@vdu.lt

 Issues and Cases of Degrowth in Tourism (ed. K. Andriotis)
DOI: 10.1079/9781789245073.0009

urgent and aggressive action. We have rung the alarm bell loud and clear' (WHO, 2020a). The worldwide pandemic situation led us to rethink our understanding about human rights as well. As human rights are interdependent, indivisible and interrelated, so this means that violating the right to health may often impair the enjoyment of other human rights (Office of the United Nations High Commissioner for Human Rights and World Health Organization, 2008).

Thus, in debating freedom of movement and degrowth, the following discussion points are opened up: How is degrowth interpreted in modern tourism? How is freedom of movement interrelated with overtourism and sustainability challenges? How are degrowth and freedom of movement related to the global health crisis? The aim of this chapter is to discuss freedom of movement and degrowth, analysing various scientific and legal sources. The chapter is organized as follows: a discussion of degrowth interpretations and modern tourism; freedom of movement's interrelationship with overtourism and sustainability challenges; and freedom of movement, degrowth and the global health crisis. Concluding insights follow.

9.2 Degrowth Interpretations and Modern Tourism

It can be noted that the magic of degrowth lies in its ability to energize and connect remarkably heterodox thinking and surprisingly heterogeneous action, as Paulson (2017) points out, including that of non-dominant groups. Such a transition is considered necessary as well as desirable. According to Gerber (2020), degrowth points to a radical political and economic transformation that results in smaller and more equitable social metabolisms. Degrowth ideas are old and originated in many parts of the world (Giacomo *et al.*, 2014). An essential approach to degrowth is the possibility to live well in egalitarian societies with much fewer resources (Kallis, 2018; Schmelzer and Vetter, 2019). According to Brossmann and Islar (2019), degrowth is a frame connecting multiple lines of thought and strategies by inviting proposals for a desired socio-ecological future motivated by social justice, democracy, respect for nature and its intrinsic value, meaning of life, wellbeing and anti-utilitarianism. Thus, as Brossmann and Islar point out, *living degrowth* – conceptualized as a transformative endeavour with certain subjective, performative, collective and structural aspects – is related in our model to current problems and imagined futures. It appears as a bridge between the two, being both 1) a response to current problems of social inequality and ecological destruction and 2) an attempt to realize imagined futures of socio-ecological sustainability.

When we discuss degrowth and freedom of movement interpretations in modern tourism it is appropriate to analyse the relevant ideas. As Renkert (2019) notices, the tourism industry is not a homogeneous entity, but is instead characterized by diverse forms of distinct 'tourisms'. According to Parrique *et al.* (2019), 'decoupling' can be discussed, taking into consideration different geographical perimeters. Adityanandana and Gerber (2019) discuss 'post-growth', contributing to emerging debates on post-growth thinking from the 'Global South' and to the radical critiques of the tourism industry in developing regions. Higgins-Desbiolles *et al.* (2019) present a conceptual consideration of issues of degrowth in tourism, examining current tensions in international mobility, and argue that just and sustainable degrowth will require greater attention to equity. Their analysis suggests that essential to such an agenda is redefining tourism to focus on the rights of local communities and a rebuilding of the social capacities of tourism. Andriotis (2014, 2018) notes the importance of the degrowth movement's guiding philosophy, which is directed towards sustainable change, contributing to a tourism society of degrowth. Higgins-Desbiolles *et al.* (2019) argue for the redefinition of tourism in order to place the rights of local communities above the rights of tourists for holidays and the rights of tourism corporates to make profits. According to Paz (2017), we now live in the midst of a massive global crisis of mobility.

Therefore, Rodríguez-Labajos *et al.* (2019) systematically evaluate both the main tensions and possible analogies between the actions of environmental justice organizations in the Global South and the propositions of the degrowth movement. Latouche (2009) has expounded eight principles for a serene and convivial journey: re-evaluate, reconceptualize, restructure, redistribute, relocalize, reduce, re-use and recycle. Accordingly, Milano *et al.* (2019) explore the evolution of the tourism degrowth discourse among social movement activists in Barcelona, and in particular where this is related to claims associated with overtourism and the extent to which this might be influencing a paradigm shift from 'tourism growth' to 'tourism degrowth'.

As a philosophical concept and movement, as Andriotis (2014) notices, degrowth is revolutionary and anti-capitalist; it is focused on sustainable change and results from an interest in locality and place, small and medium-sized enterprises, job creation and reduction of working hours, ecology and quality of life, reduction of tourism activities, carbon reduction in the transport sector, a change in production and consumption patterns and the high priority of the travel experience. According to Fletcher *et al.* (2019), to seriously pursue degrowth at both global and (most) national levels would, therefore, likely require drastic transformation of the tourism industry and its metabolism. Since tourism degrowth is a proposal, as Fletcher *et al.* (2019) argue, that aims to bring social movements and research/praxis together, scholars will have to pay particular attention to and cooperate with social movements to make another type of tourism possible within planetary boundaries. In the context of holistic transformation, including in work and public places, as Schneider and Nelson (2018) comment, residential change for degrowth will flourish and inform a broader transition to degrowth.

Thus, as mentioned previously, degrowth has extended and at the same time enriched research in the sphere of freedom of movement, in which it plays an active role which is increasingly conducted against a backdrop of resistance to progress. Still, numerous avenues deserve to be explored, including differences between tourism zones, networks, various ideas of free movement, overtourism transfers, global health crises, sustainability, politicization, etc.

9.3 Freedom of Movement's Interrelation with Overtourism and Sustainability

According to Andriotis (2014), sources of degrowth can be traced back to ancient Greece. Thus, when trying to discuss how freedom of movement is interrelated with overtourism and sustainability challenges, it seems important to have a short look at the historical view of the right of freedom of movement. This right – including the right to leave and return to one's own country – has its origins in ancient philosophy and natural law, as McAdam (2011). Articulating the vision of Socrates, Plato wrote: 'we further proclaim to any Athenian by the liberty which we allow him, that if he does not like us when he has become of age and has seen the ways of the city, and made our acquaintance, he may go where he pleases and take his goods with him. None of us laws will forbid him or interfere with him. Anyone who does not like us and the city, and who wants to emigrate to a colony or to any other city, may go where he likes, retaining his property' (Crito, 51d–e, cit. McAdam, 2011). According to McAdam, some of the first to write about the right to free movement were lawyers setting out the principles of the 'law of nations' (international law) and the writings of Spaniard Francisco de Vitoria (1492–1546) and Dutchman Hugo Grotius (1583–1645) had enormous influence on the development of such law. Summarizing and endorsing Grotius's conclusions in his 1609 work, *Mare Liberum* (*Freedom of the Seas*), McAdam points out that the safeguarding of free passage over land and waters was required by law for those 'who, for legitimate reasons, have need to cross over them; as, for instance, if a people has been forced to leave its own territories and is seeking unoccupied lands or desires to carry on commerce with a distant people, or is even seeking to recover by just war what belongs to it' (Hugo Grotius, cit. McAdam, 2011). During the French Revolution, as McAdam emphasizes, the notion of freedom of movement and the right to leave were construed as part of a broader right to liberty, whereas a right to free movement was not consistently included in the rights declarations proposed during World War II and the immediate post-war period. By 1948, the notion of a right to leave and to return to one's country was expressed as a fundamental human right worthy of recognition in the world's first universal human rights instrument. According to McAdam, the right to leave any country is not only a freestanding right but is reinforced by, and gives meaning to, other human rights. However, the right to leave a country is now a fundamental principle of human rights treaty law, but it is not an absolute right.

At first glance one might think that the difference is that freedom of movement is only fully achieved where there are no more borders (Human

Rights Committee, 1999), while admission claims imply the existence of a border, as Bauböck (2009) points out. Yet this misunderstands the nature of borders. Where free movement has become a universal right for anybody residing within the jurisdictions involved, it becomes a liberty in the sense that Wesley Hohfeld refers to as a privilege (Hohfeld, 1919). As Bauböck notices, its correlate is not a state duty but a 'no right', i.e. the absence of a right to control exit or entry; by contrast, where freedom of movement is exercised at a border that is simultaneously used for immigration control, enjoying the liberty does not imply the full absence of control but a right to be exempted.

The association between free movement and fundamental rights has a long pedigree in case law. As de Cecco (2014) points out, the term fundamental right was already present in the preamble to Regulation 1612/68, which states in its third principle: 'Freedom of movement constitutes a fundamental right of workers and their families.' Taylor (2020) notices that with clear emphasis on the self-autonomy of the individual and in common with the freedoms of opinion and of expression, liberty of movement is 'an indispensable condition for the free development of a person'. So, what is the right to freedom of movement? Under Directive 2004/38/EC of the European Parliament (2004) 'citizenship of the Union confers on every citizen of the Union a primary and individual right to move and reside freely within the territory of the Member States, subject to the limitations and conditions laid down in the Treaty and to the measures adopted to give it effect.'

The free movement of persons constitutes one of the fundamental freedoms of the EU internal market, which comprises an area without internal frontiers, in which freedom is ensured in accordance with the provisions of the Treaty. The right to freedom of movement includes the right to move freely within a country for those who are lawfully present, the right to leave any country and the right to enter a country of which you are a citizen. The right may be restricted in certain circumstances. The term freedom of movement is used by participating states of the the Organization for Security and Cooperation in Europe (OSCE) to describe a wide range of topics that concern the right of all people to leave and return to their state, and to move freely within the borders of their own state. It also covers the rights of foreigners to unrestricted movement within state borders and the facilitation of freer cross-border movement and contacts among people, institutions and organizations in participating states (Organization for Security and Cooperation in Europe, 2020). The human right to freedom of movement, like the right to health, is fundamental. In addition to being present in numerous national laws, it is protected through the United Nations (1948) and the International Covenant on Civil and Political Rights (ICCPR) (1976). Article 12 of the ICCPR provides that 'everyone lawfully within the territory of a State shall, within that territory, have the right to liberty of movement'. It further states that 'everyone shall be free to leave any country, including his own' (International Covenant on Civil and Political Rights, 1976). Unlike the right to health, the full realization of the right to freedom of movement must be achieved immediately. The right to free movement, as contained in the ICCPR, has been authoritatively interpreted by the Human Rights Committee in General Comment 27 as placing a duty on the state of residence or nationality to not unduly restrict the internal movement of persons lawfully within the country, or to place barriers on their leaving the country. Critically, the right to freedom of movement does not in any significant way impose duties on the receiving states.

When we discuss freedom of movement in the context of overtourism and sustainability we should mention other related rights such as the right to travel, local population rights, rights of travellers, right to privacy, right to a safe and clean environment, etc. As Perkumienė and Pranskūnienė (2019) propose, the relationship between the rights of travellers and the local population's right to live where they please is closely related to the concept of freedom of movement in the context of overtourism. After analysis of legal acts related to the rights of travellers and residents' rights in the context of overtourism, it can be noted that the rights of travellers are clearer and better protected, and the interpretation of the rights of the local population is too broad and not as effectively expressed. As for the legal relationship between the traveller and the local resident living in an epicentre of overtourism, it is important to conclude that overtourism is associated with the fact that the rights of travellers, who are tourists moving for entertainment and consumption purposes, are not equivalent to the residents' rights – both local residents and those people who have changed their place of residence.

In recent years, as Valdivielso and Moranta (2019) point out, overtourism and tourism gentrification have been a central feature of public debate in the Balearic Islands, one of the most touristified regions on the planet. The authors note that the debate over tourism degrowth concerns not just a discrepancy between different approaches to deal with overtourism, tourism pressure and gentrification, but is also a social struggle where the discourse over tourism degrowth has helped new democratic political subjects to coalesce and organize in civil society.

In the context of sustainable tourism (Perkumienė and Pranskūnienė, 2019), we should definitely talk about balancing the right to travel and the right to live. Panzer-Krause (2019) argues that sustainability networks support rural areas' efforts for change; however, degrowth strategists need to be represented as key actors of communication and in joint activities in the long term to realize a clear shift away from business as usual. Although social movements do not reject tourism, Navarro-Jurado *et al.* (2019) call for greater control over its impact, denounce unlimited growth, overtourism and the loss of urban quality of life. The degrowth and sustainability movements certainly share common ground, as Perryer (2019) notices, particularly with regards to reducing consumption: taking fewer flights, eschewing fast fashion, purchasing less packaged food and committing to growing produce at home are all examples of mutual initiatives. Degrowthers, however, recognize that these individual changes, while valuable in contributing to wider environmental goals, will not achieve the sort of systemic change that they believe is necessary. These movements, as Navarro-Jurado *et al.* (2019) point out, advocate a lifestyle linked to the everyday space, which they believe is threatened by excessive urban tourism growth, being a symptom of the need to devise a proposal using the principles of degrowth.

Ensuring another related human right to a clean, healthy and safe natural environment and protecting the environment from negative impacts is one of the most pressing issues, from both socioeconomic and legal/political perspectives. As Perkumienė *et al.* (2020) note, these issues depend on the recognition, regulation and application of preventive environmental methods or measures seeking a harmonized legal framework at the national and international level. Notably, in the current period of technical progress, the environment is affected by various negative factors, such as pollution, climate change and nuclear dangers that threaten not only human life and health but also humanity as a whole. These negative outcomes are global issues and require both national and international unified responses and solutions. In the scientific literature and legal documentation, the concept of a clean environment is not well established, and the most commonly used terms are the 'right to a healthy and clean environment' and the 'right to a healthy and safe environment'. In addition, the debate, as Gascón (2019) notices, is about how to manage the conflict between rights: the right of the citizen as a tourist against the rights of the citizen as a resident of a territory or as a worker; the right of the tourist to travel the planet against the rights aimed at ensuring the resilience of ecosystems.

Thus, legal instruments aim to maximize the assurance of the implementation of persons' rights to free movement in the context of overtourism and sustainability challenges. This area remains problematic, especially in international law. In this situation, the question is whether the implementation of legal norms could be enhanced between freedom of movement interrelated with overtourism and sustainability and degrowth, and freedom of movement related to the global health crisis. What are the possible solutions?

9.4 Freedom of Movement, Degrowth and the Global Health Crisis

The human rights to health and freedom of movement are fundamental and universal, and both create immediate obligations on states. The existence of these two rights in the context of managing health worker migration, particularly recommendations which limit active recruitment or promote self-sufficiency policies as part of an ethical code of practice, does not compromise a state's ability to honour both rights. Even where a conflict might arise in the fulfilment of these two human rights, as perhaps in the case of temporary domestic government bonding of health workers, restriction to the freedom of movement could be permitted under the exception provided for public health. In sum, the tension between the two internationally accepted human rights remains in large part that of perception rather than one strictly dictated by human rights law. (Honouring the Human Rights to Health and Freedom of Movement Realizing Rights, Global Policy Advisory Council Secretariat, Health Worker Migration Initiative May 7, 2009). International human rights law guarantees everyone the right to the highest attainable standard of health and obligates governments to take steps to prevent threats to public health and to provide medical care to those who need it. Human rights law also recognizes that in the context of serious public health threats and public emergencies threatening the life of the nation, restrictions on some rights can be justified when they have a legal basis, are strictly necessary based on scientific evidence, are neither arbitrary nor discriminatory in application, are of limited duration, are respectful of human dignity, subject to review and proportionate to achieve the objective (WHO, 2020a).

Are public health and freedom of movement related? Is this a new phenomenon? It is interesting to note that in July 1979 an outbreak of influenza on a commercial airliner in Alaska was reported. Fricker and Steffen (2008) explain that while being unable to take off, the aircraft ventilation system was shut down. Within 72 hours, 72% of the 54 passengers had developed symptoms of influenza, the attack rate being higher the longer the person stayed on board. The authors remind us that this event raised widespread concern about the danger of the spread of infectious diseases during air travel, just as, historically, migration and global trade resulted in pandemic levels of plague spread. Despite the fact that tourism resulted in an increase in the number of travellers, as Fricker and Steffen point out, global mobility only rarely resulted in a huge public health impact; lately, this was the case with HIV and SARS, which would have spread at a slower pace before the jet age. Baker (2014) argues that travellers should be considered an integral part of the global surveillance network for emerging infections, as research and the knowledge gained can be used to alert the global community to the presence or susceptibility patterns of pathogens in different regions and inform strategies that can be used to control infections in developing countries. Travellers to such areas can also be forewarned and the care of those returning allowed for as required. In this highly mobile world, as Fricker and Steffen (2008) note, modern public health measures such as surveillance, early case detection and management, vaccination, informing the public and facilitated access to medical services have replaced traditional quarantine measures in cases of epidemic. Baker (2014) argues that one major lesson from the Ebola outbreak was for the world to respond much more quickly than before to epidemics. The containment and preparedness efforts dramatically limited the potential impact of Ebola on the African economy, compared to earlier worst-case scenarios. This demonstrates why all countries, developing and developed, should make investing in pandemic preparedness a top priority, as global travel, according to Baker, has evolved dramatically during the past two centuries, with ever escalating speed, distance and volume.

According to Page (2007), the most difficult part of any quarantine law is determining how much to protect individual rights and liberties while still keeping the public healthy and safe. Here we can rise a question: can the right to freedom of movement be limited? The right may be restricted, either by way of derogation under Article 4 of the ICCPR, or to protect national security, public order, public health or morals, or the rights and freedoms of others, as allowed by Article 12(3) (International Covenant on Civil and Political Rights, 1976). It should be noted that the right to free movement of persons is fundamental, but it is not absolute. Article 3 of the Declaration of Human Rights guarantees that everyone has the right to life, liberty and the security of person (United Nations, 1948). The security of all citizens and the right to life as a fundamental and indisputable right, and a safe environment, must be ensured. Under the Directive 2004/38/EC of the European Parliament (2004) 'procedural safeguards should also be specified in detail in order to ensure a high level of protection of the rights of Union citizens and their family members in the event of their being denied leave to enter or reside in another Member State, as well as to uphold the principle that any action taken by the authorities must be properly justified'. Under Article 29 the only diseases justifying measures restricting freedom of movement shall be the diseases with epidemic potential as defined by the relevant instruments of the WHO and other infectious diseases or contagious parasitic diseases if they are the subject of protection provisions applying to nationals of the host Member State (Directive 2004/38/EC of the European Parliament, 2004). For example, the USA case *Jacobson* v. *Massachusetts* is the precedent case delineating the state's broad authority to curtail individual liberty and the right of free movement in furtherance of the public good (USA case *Jacobson* v. *Massachusetts*, 197 U.S. 11, 1905). In this case a Swedish immigrant, Henning Jacobson, had refused compulsory vaccination for smallpox and was fined five dollars pursuant. The court found in this case that the law on compulsory vaccination had a real and substantial connection with the protection of public health, and rejected Jacobson's contention that he had been deprived of his liberty and the right of free movement under the constitution.

Chertoff (2020) agrees that the right to travel is not absolute and it can be limited. Whether the border is land, an international airport or a marine port, US customs and border control may refuse entry to non-residents and place conditions like quarantine on returning US residents. The authority to address contagious diseases introduced into the US was first tested in 1963 and later in a 2007 case. Both involved individuals who had travelled abroad, and upon returning to the USA were placed in federal quarantine. In the 1963 case, a habeas petition was denied after the American traveller, Ellen Siegel, returned from Sweden, where there had been a smallpox outbreak, and a fourteen-day quarantine was upheld (US case ex rel. Siegel v. Shinnick, 219 F. Supp. 789, 1963). The 2007 case involved an American who travelled to Europe while suffering from multi-drug resistant tuberculosis, and was isolated for a total of five days. In each case, a specific named individual was the subject of the quarantine, and there was ample evidence that the individual posed a risk of infection to others, and due process was provided during the quarantine period (Chertoff, 2020).

An analysis of the legal literature and court rulings reveals that there are also several violations related to the restriction of rights related to the protection of national security and public health. For example, court decision I/A Court H.R., Matter of the Penitentiary Complex of Curado regarding Brazil (2016) noted that Brazil had not presented a detailed plan to provide medical assistance to inmates with contagious diseases. The court directed Brazil to adopt additional urgent measures to ensure prisoners were provided with adequate medical treatment and that inmates, personnel and visitors were not exposed to contagious diseases. The court also stated that Brazil should adopt preventive health policies, taking into consideration the special needs of vulnerable inmates, including prisoners with tuberculosis and HIV. Ulrich and Mariner (2018) raise the question as to whether the USA has the legal framework to prevent or respond to the next pandemic health crisis. They point out that the world has so far faced real or potential epidemics in the form of HIV, SARS, Zika and Ebola. With tourism and exploration expanding into forests, jungles and caves, the next pandemic could be sparked by an unsuspecting tourist visiting a part of the world that humans have rarely seen (Ulrich and Mariner, 2018). Mao *et al.* (2010) argue that tourism demand is particularly sensitive to security and health concerns and the industry is highly susceptible to changes in the international political situation, to natural disasters and to epidemics. They notice that the WHO issued a global alert with respect to SARS on 12 March 2003, as the disease continued to spread rapidly. This alert was followed on 15 March by a recommendation that emergency travel restrictions be imposed, advice that has been extremely rare in the history of the WHO. Besides the direct impact on global economic activity, as Mao *et al.* emphasize, the fear of infection and the quarantine measures imposed by governments around the world led many people to cancel or change their travel plans. The SARS epidemic, therefore, had a severe negative impact on global travel and the tourism industry.

We also can raise the same question for the international legal framework to prevent a global health crisis. However, the COVID-19 pandemic reveals gaps in the international legal framework for prevention and management of such crises. How are public health and travel related in these situations? Ying *et al.* (2020) point out that on 31 January 2020, the WHO declared China's coronavirus outbreak a public health emergency of international concern, only the sixth of its kind (ABC News, 2020). On 8 February 2020, the National Health Commission of the People's Republic of China (2020) officially renamed the new coronavirus 'novel coronavirus pneumonia' (NCP). The disease originated in Wuhan, Hubei Province, China, and has been classified as a global epidemic. Therefore, we have no doubts about the current situation with COVID-19. On 17 March 2020 EU Member States agreed on coordinated action at external borders based on the recommendation by the Commission to restrict non-essential travel for a specific period which has been extended a number of times (European Commission, 2020). On 11 June 2020, the Commission adopted a Communication which set out an approach to progressively lift the restrictions. So, the global health crisis and free movement become practically incompatible. In the face of this unprecedented pandemic situation that has claimed many lives, there is no point in discussing the importance of freedom of movement or the right to travel. To manage the crisis, we must restrict the free movement of people. We can argue that law cannot guarantee comprehensive solutions – it only shows the direction where we must go.

On the one hand, it refers, as Toebes (2015) points out, to the right to health by stressing the importance of the need to 'protect health workers from infection', limit the use of quarantines, address the gender dimensions of the outbreak and ensure any security forces responding to the crisis respect basic rights; on the other hand, it states that 'donor governments, through international assistance and cooperation, should help these governments fulfil the right to health and efforts to address the epidemic's broader impact'. Davey (2020) points out that SARS-Cov-2 has put 'degrowth' in a new perspective and on a different and a necessarily more rapid timescale. Previously, when arguing the case for degrowth, Davey, 2020 had been thinking and writing about a contraction of economic activity to be achieved in years or decades, but now the global pandemic requires action to curb dangerous connectivity and thus dangerous activities in a matter of days, weeks and months. It means staying at home – and thus not using fossil fuel powered transport (including public transport infrastructures and aeroplanes); it means minimal shopping – only for essentials. Self-isolation and distancing are widespread. This is set to bring the economy down and is already doing so. It is not an ideal model for degrowth.

According to Maxwell *et al.* (2020), the metaphor of war is often used in the infectious diseases field, with its interspecies fight for survival. As the battle against COVID-19 ramps up worldwide, it is imperative that the entire global community joins together in solidarity, applies the hard-fought lessons of this and prior epidemics, and moves rapidly to implement proven public health and IPC principles to turn the tide against this foe.

It should be noted that in the current global health crisis three rights are at the front line: 1) the right to life and duty to protect life; 2) the right to health and access to health care; 3) the right to freedom of movement (COVID-19 and National Health Commission of the People's Republic of China, 2020). As discussed above, the application of the right to life reminds us that human lives must be protected, including by addressing the general conditions of society that pose an imminent threat to life. The right to health is inseparable from the right to life. We are currently witnessing a situation where COVID-19 tests the limited capacity of states to defend the right to health. Everything seems to be being done to guarantee people's rights to life and health, but it has not been possible to find rational and comprehensive solutions as yet. Controlling the virus and defending the right to life means breaking the chain of infection: people need to stop moving. Thus, in this case, as we see, the most common public health measure against COVID-19 is the restriction of freedom of movement: staying at home. This measure is a practical and necessary way to stop the spread of viruses, prevent pressure on health care and thus save lives (United Nations, 2020). It is clear and seems self-evident that in this pandemic it is perfectly justified that in order to preserve one right, it is necessary to constrain another one. And there is not even a question of which of these rights are more important. The right to free movement in this critical situation becomes less important.

Applying critical health literacy, as Abel and McQueen (2020) notice, has never been more needed than in these days when an infectious disease crisis emerges at a time of information excess and high expectations of controlling health. These advanced expectations meet with another unique condition: never in human history has there been such an abundance of health information available from numerous more or less trustworthy sources. Brunette and Nemhauser (2020) propose that, because most common travel-related infections have short incubation periods, a majority of ill travellers will seek medical attention within a month of return from their destination. According to Frost *et al.* (2019), increasing long-distance travel and the increasing popularity of undertaking medical procedures abroad have increased the risk of highly resistant germs being transferred across borders. Schwerdtle *et al.* (2019) notice that humans have a long history of mobility on a spectrum from voluntary migration to forced displacement in response to social, political and environmental change. Chohan (2020) notes that the coronavirus outbreak has left governments, markets and civil society reeling from disruptions and damage that will heal at differing intervals and to differing degrees around the world. Strielkowski (2020) emphasizes that COVID-19 means bad news for international travel and tourism. Smart quarantine tools might be used to mark healthy and risk-free travellers and such technological solutions would help to reduce the risk of spreading COVID-19 before an effective vaccine is discovered, without compromising international tourism and travel. By 2050, as Holubar (2020) believes, infections due to multidrug-resistant organisms are projected to become the leading cause of death worldwide, outpacing cancer, diabetes and motor vehicle accidents. We live in an increasingly connected world; the rise and spread of antimicrobial resistance (AMR) reminds us that multidrug-resistant organisms do not respect geographic borders, so the emergence of such an organism in one part of the world threatens the entire international community and must be treated as a global public health emergency.

We should note that international human rights law guarantees everyone the right to the highest attainable standard of health and obliges governments to take measures to prevent threats to public health and to provide medical care to those who need it. As noted above, human rights law also recognizes that in the event of a serious threat to public health and emergencies endangering the life of a nation, restrictions on certain rights, such as the right to free movement of persons, may be justified when they have a legal basis, are strictly necessary, are based on scientific evidence and are not arbitrary or unjustifiable. They must be of limited duration, respectful of human dignity, reviewed and proportionate (United Nations, 2020). What about degrowth here? How could degrowth be relevant? As Brossmann and Islar (2019) notice, degrowth develops an alternative imaginary of an economy of care and commons, and endorses new forms of living and producing like eco-communities and cooperatives as well as policies for work-sharing or basic and maximum income. A deeper understanding of the less than rational and less than physical ways in which we create our world experience is needed to move the planet in a more positive direction (Galvani *et al.*, 2020). As mentioned previously, law cannot guarantee comprehensive solutions – it only shows the direction where we should go, so the biggest challenge here is seeking a balance between fundamental rights and the alternative imaginary of an economy.

9.5 Concluding Insights

This chapter has debated freedom of movement and degrowth, and has tried to addess how degrowth is currently interpreted, how freedom of movement is understood and how freedom of movement and degrowth are related in terms of the global health crisis. As Higgins-Desbiolles (2020) notices, the arrival of COVID-19 has been viewed as a watershed moment, and in terms of tourism the unthinkable has happened. While previously the industry was grappling with concerns of overtourism, we have suddenly been catapulted into a collapse of the entire sector. Galvani *et al.* (2020) point out that COVID-19 is a once-in-a-lifetime experience for everyone on planet Earth, as there has never been an event in the recorded history of modern humankind that has affected everyone on the planet to the same degree. Beyond the immediate health crisis, COVID-19 is essentially a crisis of economized societies rooted in the growth paradigm (Gretzel *et al.*, 2020; Ötsch, 2020).

It seems important to point out that the pandemic situation has led to more and more initiatives uniting legal instruments and rules. For example, the European Commission launched 'Re-open EU', a web platform that contains essential information for allowing a safe relaunch of free movement and tourism across Europe (European Commission, 2020), and the COVID-19 Law Laboratory initiative gathers and shares legal documents from over 190 countries across the world to help states establish and implement strong legal frameworks to manage the pandemic (WHO, 2020b). These and other emerging initiatives have led Ulrich and Mariner (2018) to pose the question: would it be possible to have a worldwide legal framework to prevent or respond to a global pandemic health crisis?

Overtourism and undertourism are inevitable in an era characterized by globalization, sophistication and emancipation, but as Gowreesunkar and Vo Thanh (2020) note, they are also the outcome of mismanaged destinations, poor planning and lack of coordination. Both are causes of the destruction of destinations and both have to share responsibility. After all, different stakeholders nurture contrasting and conflicting interests. As Ötsch (2020) comments, people talk about a society that turns a natural phenomenon – the virus – into a social crisis, and a more positive approach requires that we recognize the positive moments of the coronavirus shock, emphasize them and bring them to the fore.

Degrowth can be considered not only a contemporary but also a controversial approach. Its main features make it highly differentiated from all other forms of tourism development. Gretzel *et al.* (2020) point out that despite emergent questioning of the growth paradigm for tourism, present before the crisis, much of the current discourse concerning tourism during the pandemic is coloured by the notion that tourism is a 'demand-driven' economic activity that needs to be 'stimulated' so that it can 'grow' again. Both science and technology are regarded as means to serving this end, rather than transformative forces. As Gretzel *et al.* argue, researchers must come out from their own silos to build bridges that lead to fruitful collaboration.

We must all rethink the meaning of freedom of movement in tourism, thus opening up broader discussion for the future development of tourism research. As we try to give birth to a new era in the aftermath of COVID-19, as Higgins-Desbiolles (2020) notices, it is our task as concerned scholars in tourism to offer our contributions to imagine ways tourism can be developed to enable humans to thrive and to engender ecological recovery. We understand that to comprehensively answer the questions posed in this chapter is difficult or even impossible, but we seek to pay attention to these challenging problems and to invite contributions to further debates.

References

ABC News (2020) Coronavirus is a Public Health Emergency. World Health Organization declares. ABC News. Available at: https://www.abc.net.au/news/2020-01-31/who-coronavirus-global-emergency/11916276 (accessed 15 July 2020).

Abel, T. and McQueen, D. (2020) Critical health literacy and the COVID-19 crisis. *Health Promotion International* 2020, 1–2.

Adityanandana, M. and Gerber, J.-F. (2019) Post-growth in the tropics? Contestations over *Tri Hita Karana* and a tourism megaproject in Bali. *Journal of Sustainable Tourism* 27(12), 1839–1856. DOI: 10.1080/09669582.2019.1666857.

Andriotis, K. (2014) Tourism development and the degrowth paradigm. *Proceedings of the 2nd Belgrade International Tourism Conference (BITCO 2014)*, Thematic Tourism in a Global Environment: Advantages, Challenges and Future Developments, Belgrade, Serbia, 27–29 March 2014, pp. 37–945.

Andriotis, K. (2018) *Degrowth in Tourism: Conceptual, Theoretical and Philosophical Issues.* CAB International, Wallingford, UK.

Baker, D. (2014) Tourism and the health effects of infectious diseases: are there potential risks for tourists? *International Journal of Safety and Security in Tourism/Hospitality*, 1–17.

Bauböck, R. (2009) Global justice, freedom of movement and democratic citizenship. *European Journal of Sociology* 50(1), 1–31. DOI: 10.1017/S000397560900040X.

Brossmann, J. and Islar, M. (2019) Living degrowth? Investigating degrowth practices through performative methods. *Sustainability Science.*

Brunette, G.W. and Nemhauser, J.B. (2020) *Posttravel Evaluation.* CDC Yellow Book 2020: Health Information for International Travel.

Chertoff, M. (2020) The right to travel and national quarantines: coronavirus tests the limits. Available at: https://www.law.georgetown.edu/salpal/the-right-to-travel-and-national-quarantines-coronavirus-tests-the-limits/ (accessed 25 July 2020).

Chohan, U.W. (2020) A Post-Coronavirus World: 7 Points of Discussion for a New Political Economy. *CASS Working Papers on Economics & National Affairs* No. EC015UC, March 2020. Available at: SSRN: http://dx.doi.org/10.2139/ssrn.3557738 (accessed 12 May 2020).

Davey, B. (2020) Coronavirus, degrowth and self isolation. Available at: https://www.resilience.org/stories/2020-03-18/coronavirus-degrowth-and-self-isolation/ (accessed 10 July 2020).

de Cecco, F. (2014) Fundamental freedoms, fundamental rights, and the scope of free movement law. *German Law Journal* 15(3), 383–406. DOI: 10.1017/S2071832200018964.

Directive 2004/38/EC of the European Parliament (2004) Directive 2004/38/EC of the European Parliament and of the Council of 29 April 2004 on the right of citizens of the Union and their family members to move and reside freely within the territory of the member states amending regulation (EEC) NO 1612/68 and repealing directives 64/221/EEC, 68/360/EEC, 72/194/EEC, 73/148/EEC, 75/34/EEC, 75/35/EEC, 90/364/EEC, 90/365/EEC and 93/96/EEC. Available at: https://eur-lex.europa.eu/legal-content/EN/TXT/?uri=CELEX:32004L0038 (accessed 10 May 2020).

European Commission (2020) Travel during the coronavirus pandemic. European Commission. Available at: https://ec.europa.eu/info/live-work-travel-eu/health/coronavirus-response/travel-and-transportation-during-coronavirus-pandemic/travel-and-eu-during-pandemic_en (accessed 25 July 2020).

Fletcher, R., Murray Mas, I., Blanco-Romero, A. and Blázquez-Salom, M. (2019) Tourism and degrowth: an emerging agenda for research and praxis. *Journal of Sustainable Tourism* 27(12), 1743–1763. DOI: 10.1080/09669582.2019.1679822.

Fricker, M. and Steffen, R. (2008) Travel and public health. *Journal of Infection and Public Health* 1(2), 72–77. DOI: 10.1016/j.jiph.2008.10.005.

Frost, I., Van Boeckel, T.P., Pires, J., Craig, J. and Laxminarayan, R. (2019) Global geographic trends in antimicrobial resistance: the role of international travel. *Journal of Travel Medicine* 26(8). DOI: 10.1093/jtm/taz036.

Galvani, A., Lew, A.A. and Perez, M.S. (2020) COVID-19 is expanding global consciousness and the sustainability of travel and tourism. *Tourism Geographies* 22(3), 567–576. DOI: 10.1080/14616688.2020.1760924.

Gascón, J. (2019) Tourism as a right: a "frivolous claim" against degrowth? *Journal of Sustainable Tourism* 27(12), 1825–1838. DOI: 10.1080/09669582.2019.1666858.

Gerber, J.-F. (2020) Degrowth and critical agrarian studies. *The Journal of Peasant Studies* 47(2), 235–264. DOI: 10.1080/03066150.2019.1695601.

Giacomo, D., Demaria, F. and Kallis, G. (2014) *Degrowth: A Vocabulary for a New Era*. Routledge.

Gorz, A. (1980) Ecology and freedom. In: *Ecology as Politics*. Black Rose Books, Québec.

Gowreesunkar, V.G. and Vo Thanh, T. (2020) Between overtourism and under-tourism: impacts, implications, and probable solutions. In: Séraphin, H., Gladkikh, T. and Vo Thanh, T. (eds) *Overtourism*. Palgrave Macmillan, Cham.

Gretzel, U., Fuchs, M., Baggio, R., Hoepken, W., Law, R. *et al.* (2020) e-Tourism beyond COVID-19: a call for transformative research. *Information Technology & Tourism* 22(2), 187–203. DOI: 10.1007/s40558-020-00181-3.

Higgins-Desbiolles, F. (2020) Socialising tourism for social and ecological justice after COVID-19. *Tourism Geographies* 22(3), 610–623. DOI: 10.1080/14616688.2020.1757748.

Higgins-Desbiolles, F., Carnicelli, S., Krolikowski, C., Wijesinghe, G. and Boluk, K. (2019) Degrowing tourism: rethinking tourism. *Journal of Sustainable Tourism* 27(12), 1926–1944. DOI: 10.1080/09669582.2019.1601732.

Hohfeld, W. (1919) *Fundamental Legal Conceptions as Applied in Juridical Reasoning*. Yale University Press, New Haven.

Holubar, M. (2020) Antimicrobial resistance: a global public health emergency further exacerbated by international travel. *Journal of Travel Medicine* 27(1). DOI: 10.1093/jtm/taz095.

Honouring the Human Rights to Health and Freedom of Movement Realizing Rights, Global Policy Advisory Council Secretariat, Health Worker Migration Initiative May 7 (2009). Available at: https://www.who.int/workforcealliance/about/taskforces/migration/MigrationTaskForce_techbrief.pdf (accessed 10 July 2020).

Human Rights Committee (1999) General Comment No. 27 on the freedom of movement, 1999 U.N. doc CCPR/C/21/Rev.1/Add.9. University of Minnesota, Human Rights Library. Available at: http://hrlibrary.umn.edu/gencomm/hrcom27.htm (accessed 12 March 2020).

International Covenant on Civil and Political Rights (1976) International covenant on civil and political rights (ICCPR), adopted and opened for signature, ratification and accession by G.A. Res. 2200A (XXI) of 16 December 1966, entered into force 23 March 1976, in accordance with article 49, art. 12. Available at: http://www.ohchr.org/english/law/ccpr.htm (accessed 12 January 2020).

Kallis, G. (2018) *Degrowth (The economy: Key ideas)*. Agenda Publishing, New York, NY.

Latouche, S. (2009) *Farewell to growth*. Polity Press, Malden.

Mao, C.-K., Ding, C.G. and Lee, H.-Y. (2010) Post-SARS tourist arrival recovery patterns: an analysis based on a catastrophe theory. *Tourism Management* 31(6), 855–861. DOI: 10.1016/j.tourman.2009.09.003.

Maxwell, D.N., Perl, T.M. and Cutrell, J.B. (2020) "The Art of War" in the Era of Coronavirus Disease 2019 (COVID-19). *Clinical Infectious Diseases* XX. DOI: 10.1093/cid/ciaa229.

McAdam, J. (2011) An intellectual history of freedom of movement in international law: the right to leave as a personal liberty [online]. *Melbourne Journal of International Law* 12(1), 27–56.

Milano, C., Novelli, M. and Cheer, J.M. (2019) Overtourism and degrowth: a social movements perspective. *Journal of Sustainable Tourism* 27(12), 1857–1875. DOI: 10.1080/09669582.2019.1650054.

National Health Commission of the People's Republic of China (2020) Notice of the National Health Commission on the provisional naming of the new coronavirus pneumonia. National Health Commission of the People's Republic of China. Available at: http://www.nhc.gov.cn/xcs/zhengcwj/202002/18c1bb43965a4492907957875de02ae7.shtml (accessed 26 July 2020).

Navarro-Jurado, E., Romero-Padilla, Y., Romero-Martínez, J.M., Serrano-Muñoz, E., Habegger, S. *et al.* (2019) Growth machines and social movements in mature tourist destinations Costa del Sol-Málaga. *Journal of Sustainable Tourism* 27(12), 1786–1803. DOI: 10.1080/09669582.2019.1677676.

Office of the United Nations High Commissioner for Human Rights, World Health Organization (2008) The Right to Health. United Nations Human Rights. Office of the High Commissioner. Available at: https://www.ohchr.org/Documents/Publications/Factsheet31.pdf (accessed 16 May 2020).

Organization for Security and Co-operation in Europe (2020) Freedom of Movement. Available at: https://www.osce.org/odihr/freedom-of-movement (accessed 10 July 2020).

Ötsch, W. (2020) What type of crisis is this? The coronavirus crisis as a crisis of the economicised Society. *Working Paper Serie*, No. 57. Cusanus Hochschule für Gesellschaftsgestaltung, Institute für Ökonomie und für Philosophie, Bernkastel-Kues.

Page, E.M. (2007) Balancing individual rights and public health safety during quarantine: the U.S. and Canada. Case Western reserve. *Journal of International Law* 38(3).

Panzer-Krause, S. (2019) Networking towards sustainable tourism: innovations between green growth and degrowth strategies. *Regional Studies* 53(7), 927–938. DOI: 10.1080/00343404.2018.1508873.

Parrique, T., Barth, J., Briens, F., Kerschner, C., Kraus-Polk, A. *et al.* (2019) *Decoupling debunked: Evidence and arguments against green growth as a sole strategy for sustainability.* European Environmental Bureau, Bruxelles.

Paulson, S. (2017) Degrowth, culture and power. *Special Section of the Journal of Political Ecology* 24, 426–448.

Paz, M. (2017) The incomplete right to freedom of movement. *AJIL Unbound* 111, 514–518. DOI: 10.1017/aju.2018.23.

Perkumienė, D. and Pranskūnienė, R. (2019) Overtourism: between the right to travel and residents' rights. *Sustainability* 11(7), 2138. DOI: 10.3390/su11072138.

Perkumienė, D., Pranskūnienė, R., Vienažindienė, M. and Grigienė, J. (2020) The right to a clean environment: considering green logistics and sustainable tourism. *International Journal of Environmental Research and Public Health* 17(9), 3254. DOI: 10.3390/ijerph17093254.

Perryer, S. (2019) Cutting back: how the degrowth movement could save the planet. Available at: https://www.europeanceo.com/lifestyle/cutting-back-how-the-degrowth-movement-could-save-the-planet/ (accessed 16 May 2020).

Renkert, S.R. (2019) Community-owned tourism and degrowth: a case study in the Kichwa Añangu community. *Journal of Sustainable Tourism* 27(12), 1893–1908. DOI: 10.1080/09669582.2019.1660669.

Rodríguez-Labajos, B., Yánez, I., Bond, P., Greyl, L., Munguti, S. *et al.* (2019) Not so natural an alliance? Degrowth and environmental justice movements in the Global South. *Ecological Economics* 157, 175–181. DOI: 10.1016/j.ecolecon.2018.11.007.

Schmelzer, M. and Vetter, A. (2019) *Degrowth/Postwachstum zur Einführung*. Junius Verlag, Hamburg.

Schneider, F. and Nelson, A. (2018) *Housing for degrowth. Conference: Degrowth Movement: Integration before Institutionalisation.* At Free University of Brussels, Brussel.

Schwerdtle, P.N., Bowen, K., McMichael, C. and Sauerborn, R. (2019) Human mobility and health in a warming world. *Journal of Travel Medicine* 26(1). DOI: 10.1093/jtm/tay160.

Strielkowski, W. (2020) International tourism and COVID-19: recovery strategies for tourism organisations. *Preprints* 2020030445. DOI: 10.20944/preprints202003.0445.v1.

Taylor, P. (2020) Article 12: freedom of movement of the person. In: *A Commentary on the International Covenant on Civil and Political Rights: The UN Human Rights Committee's Monitoring of ICCPR Rights*. Cambridge University Press, Cambridge, pp. 325–353.

Toebes, B. (2015) Human rights and public health: towards a balanced relationship. *The International Journal of Human Rights* 19(4), 488–504. DOI: 10.1080/13642987.2015.1044814.

Ulrich, M.R. and Mariner, W.K. (2018) Quarantine and the federal Rolein epidemics. SMU law review, volume 71/2. Available at: https://scholar.smu.edu/cgi/viewcontent.cgi?article=4728&context=smulr (accessed 28 July 2020).

United Nations (1948) Universal Declaration of Human Rights, art. 13. Available at: https://www.un.org/en/universal-declaration-human-rights/ (accessed 14 January 2020).

United Nations (2020) COVID-19 and human rights: we are all in this together. United Nations. Available at: https://www.un.org/victimsofterrorism/sites/www.un.org.victimsofterrorism/files/un_-_human_rights_and_covid_april_2020.pdf (accessed 12 May 2020).

US case ex rel. Siegel v. Shinnick, 219 F. Supp. 789 (1963) Available at: https://law.justia.com/cases/federal/district-courts/FSupp/219/789/1438260/ (accessed 28 July 2020).

USA case Jacobson v. Massachusetts, 197 U.S. 11 (1905) Available at: https://supreme.justia.com/cases/federal/us/197/11/ (accessed 28 July 2020).

Valdivielso, J. and Moranta, J. (2019) The social construction of the tourism degrowth discourse in the Balearic Islands. *Journal of Sustainable Tourism* 27(12), 1876–1892. DOI: 10.1080/09669582.2019.1660670.

WHO (2020a) Director-General's opening remarks at the media briefing on COVID-19 - 11 March 2020. Available at: https://www.who.int/dg/speeches/detail/who-director-general-s-opening-remarks-at-the-media-briefing-on-covid-19-11-march-2020 (accessed 10 May 2020).

WHO (2020b) New COVID-19 law lab to provide vital legal information and support for the global COVID-19 response, 22 July 2020. Available at: https://www.who.int/news-room/detail/22-07-2020-new-covid-19-law-lab-to-provide-vital-legal-information-and-support-for-the-global-covid-19-response (accessed 26 July 2020).

Ying, T., Wang, K., Liu, X., Wen, J. and Goh, E. (2020) Rethinking game consumption in tourism: a case of the 2019 novel coronavirus pneumonia outbreak in China. *Tourism Recreation Research* 969(1), 1–6. DOI: 10.1080/02508281.2020.1743048.

10 Conclusion

Konstantinos Andriotis*

Middlesex University, London, UK

'Questioning tourism development is tantamount to challenging the current capitalist productive model and its growth imperative' (Fletcher *et al.*, 2019, p. 1753), which increases the price of land/housing, restricts the rights of the local population and contributes to processes such as globalization and climate change. As Boonstra and Joose (2013) point out, any policy aiming to halt growth in capitalist economies would be an oxymoron since restriction of growth and respect of ecological limits are incompatible with capitalism. In the light of this situation, there is a need for the abandonment of capitalism through a drastic transformation of the tourism industry (Fletcher *et al.*, 2019) and a paradigm shift away from top-down Eurocentric models of 'tourism growth' towards 'tourism degrowth' (Milano *et al.*, 2019, p. 1857). In fact, several destinations and countries have grown too much and, for them, growth is not an appropriate option any more (Hickel, 2015). As Andriotis (2018, p. 14) remarks, because of:

> unlimited growth and the associated negative effects of overconsumption and overproduction, our world today faces a hostile climate, depleted resources, the destruction of habitats, as well as severe economic crisis. Admittedly, our planet does not have the ecological and economic capacity to fulfil the dream that growth will lead to a materialistic utopia (Jackson, 2009, p. 203). So, the only choice is to find an alternative to increased growth by transforming the structures and institutions that shape the world and by articulating a more credible vision for a lasting prosperity (Jackson, 2009, p. 203).

To seriously pursue degrowth, 'limits to growth' and 'zero growth' are needed, which, in the case of overcrowded tourism destinations, would take the form of restrictions on tourist accommodation and reduced number of visitors (Mas *et al.*, 2017; Valdivielso and Moranta, 2019). This is desirable today in many rich countries that should justify their excesses and search for ways to 'catch down' to more appropriate levels of development, as well as in less developed ones that should not be regarded always as underdeveloped, but perhaps as *appropriately developed*. Thus, degrowth is an effective strategy for

*k.andriotis@mdx.ac.uk

DOI: 10.1079/9781789245073.0010

the future of most destinations and countries worldwide. As an alternative to the inherently destructive forces of 'imperialistic' mass tourism predicated on continual expansion, a radical sociopolitical and environmental transformation is required that will eventually lead to a steady state. In fact, degrowth transformation starts only through bottom-up approaches and human action, having as its ultimate goal a steady state kept within ecological limits (Boonstra and Joose, 2013), and offering an opportunity for local people to have a fair share of tourism benefits.

Despite limited past research, the increasing number of publications in the last two years, including this book, indicates that scholars have started to seriously engage tourism with the discourse of degrowth. Nevertheless, various aspects of degrowth need further investigation from a tourism perspective. This is because degrowth came to the surface only recently as a distinctive multidisciplinary academic paradigm. Taking past research negligence into consideration, this book has aimed to address the paucity of combined research on tourism and degrowth by presenting case studies on the dynamics of degrowth from different parts of the world.

Chapter 2 by Martin Fontanari, Anastasia Traskevich and Hugues Seraphin focused on the importance of destination resilience in the context of overtourism. The authors used a Delphi technique and found that overtourism requires a broader conceptual perspective that considers the idea of degrowth for mass tourism, because mass tourism expands (and multiplies) itself through globally vulnerable supply chains. At the same time, the authors argued that shaping the touristic uniqueness of destinations is based on the impacts of new partnership behaviours between all tourism participants in a region. Therefore, building a framework for destination resilience represents a central strategic solution. The authors concluded that the coherent application of the concepts of degrowth and destination resilience contributes to solving the problem of overtourism and the related negative impacts. They suggested a need to design a sociocultural tourism satellite account to measure the impact of tourism and bring native communities into a permanent self-evaluation of perceived sociocultural benefits or losses.

In Chapter 3 Ernest Cañada conducted in-depth interviews with members of three community-based tourism initiatives to explore the potentialities and limitations of community-based tourism in Central America from the perspective of a socioecological transition. Among the main findings of the study were that the three initiatives consolidated their commercial strategy, a fact that has allowed them to increase their incomes, bring economic stability and improve the living standards of their members. This strategy has also generated experiences that increase visitors' satisfaction and place the initiatives to adopt post-capitalist transition processes within the degrowth logic.

In Chapter 4, Ammalia Podlaszewska described her empirical analysis of tourist destinations in Indonesia. The study found that commoditization of locality is a tool to readjust the balance of power and reassert the views of the local community against those of developers and local authorities, and concluded that the degrowth paradigm in the Global South presents differences compared to the Global North.

In a similar vein to Milano (2017), who identified cases of 'tourismphobia' in Barcelona, and proposed structural changes to the economic model in which tourism is embedded in the city, Neil Hughes and José Mansilla's contribution (Chapter 5) adopted a multi-disciplinary framework drawing on political discourse analysis and Althusserian conceptualizations of ideology and the wider degrowth and tourism literature, to explore several issues of tourism degrowth in Spanish cities. In attempts to find alternatives to mass tourism that give priority to the environment and local community, the authors identified a three-pronged strategy aiming to shape tourism policy in Barcelona based on countering official discourse, direct action and working with institutional actors.

By using an evolutionary approach, Chapter 6 by Ammalia Podlaszewska acknowledged the substantial role of German tour operators who have gained the CSR (corporate social responsibility) certification 'TourCert' and evaluated their potential to act as change agents for degrowth-inspired tourism. Through the use of tour operators' sustainability or audit reports it was found that the TourCert tour operators follow several principles that include the sociocultural dimension of tourism, mainly close cooperation with local partners, in line with degrowth-inspired travelling. However, the study showed that the German TourCert operators represent only a small part of the whole tourism industry, and therefore their potential to engender change towards degrowth is very limited. As a result, the chapter concluded that the German CSR certification schemes can only marginally advocate and diffuse certain elements of degrowth-oriented tourism.

In Chapter 7, Pascal Mayer used the philosophical concept of heterotopia (Foucault and Miskowiec, 1986) to examine alternative travellers visiting the island of Gavdos, and reported how degrowth in tourism should be in terms of basic infrastructure, limited use of goods and services, frugality, spending cuts, slow pace and commitment over time. Mayer also identified two main differences in his study in comparison to the study of Andriotis (2013), attributed to the different methodological approaches adopted by each, as well as the timing. He concluded that because the Andriotis study took place during the Greek economic crisis, his interviewees' responses were focused against capitalism and established norms, while the present study identified other issues related to the environmental resources of the place and personal development.

In a different context, Brendan Canavan in Chapter 8 reiterated the argument, as Hall (2009), Andriotis (2018) and Castaldo (2012) had previously, that degrowth requires a moral shift to 'rightsize' tourism growth. This was concluded because in the case of the Isle of Man involuntary tourism degrowth has had extensive negative impacts on the local community. Canavan also found that while the tourism sector has been successfully replaced by other sectors that have brought financial prosperity to the island, several negative social and environmental consequences of tourism degrowth also emerged. Because of those negative effects the study found that local people were dissatisfied by the loss of tourism and the great disconnection from overseas cultures. Therefore they would like to see a return to a more touristic state. Thus the case of the Isle of Man may serve as a warning about the negative consequences which may result from involuntary tourism degrowth.

In Chapter 9, Rasa Pranskūnienė and Dalia Perkumienė asserted that all of us should rethink the meaning of freedom of movement in tourism and with their contribution attempted to open up a broader discussion for the future development of degrowth research. They concluded that scholars in the post COVID-19 era should contribute to conceptualizing ways in which tourism can benefit both human and ecological needs.

It is clear that while some may think that controlling, halting and slowing down tourism growth is simple, this is not true. Boonstra and Joose (2013) point out that true 'degrowth cannot be realised within a capitalist society, since growth is the sine qua non for capitalism' (p. 2013). Thus, degrowth as a development approach can either be used in destinations that try to build their tourism industry from scratch or those trying to move from a growth economy to a balanced one, so that tourists, hosts and local businesses use fewer environmental resources. In practice, degrowth societies are 'societies that are organized according to fundamentally different cultural, social, economic, political and technological principles than the ones organized around the growth ideology' (Büchs and Koch, 2019, p. 162). Paraphrasing Latouche (2007), the degrowth transition for western societies is possible only through a large-scale cultural reorientation that will reinforce radical changes to social structures, values and activities.

Thus, for a community to be empowered by tourism, host voices and needs should guide tourism development (Scheyvens, 1999;Chassagne and Everingham, 2019; Higgins-Desbiolles *et al.*, 2019). Moving away from top-down unsustainable tourism approaches, Latin American countries, such as Equador, have adopted alternative visions focusing on indigenous forms of good living through the increase of social and environmental wellbeing, or, in their own words, *buen vivir* that requires bottom-up community-based tourism initiatives where local communities are actually benefiting economically and socially (Scheyvens, 1999). *Buen vivir*, in the words of Chassagne and Everingham (2019) 'is based on the needs of the community, rather than ... a Eurocentric ideal of development supported by a policy of extractivism ... degrowth is a consequence of *buen vivir*, rather than the objective' (p. 1).

10.1 Concluding Remarks

Degrowth is 'a project of living meaningfully, enjoying simple pleasures, commoning, sharing and relating more with others, and working less, in more equal societies. [Its] goal is to purposefully slow things down in order to minimize harm to humans and earth systems and to reduce exploitation' (Kallis *et al.*, 2020). In effect, degrowth needs to be among the key global efforts to tackle climate change and to protect resources for future generations' basic needs and the wellbeing of people. Irrespective of this, degrowth should not be confused with the misery, forced deprivation and restrictions enforced during the COVID-19 pandemic.

On the bright side, the COVID crisis has challenged the premises of unlimited growth and the existing growth tourism model by bringing to the surface the need to consider zero-carbon goals as against development models which have produced undesirable features such as significant leakage and repatriation of profits, outside intervention, high levels of foreign ownership, overconsumption, overproduction and unequal spatial development (Andriotis, 2002, 2005, 2008, 2011; Das and Sharma, 2009; Chen and Var, 2010; Lacher and Nepal, 2010; Faccioli, 2011; Monterrubio and Andriotis, 2014; Eusébio *et al.*, 2016; Mbaiwa *et al.*, 2018). Thus, the COVID-19 outbreak may lead to a critical reconsideration of the global volume growth model for tourism (Gössling *et al.*, 2020, p. 13). In the words of Gössling *et al.* (2020) 'the COVID-19 crisis should be seen as an opportunity to critically reconsider tourism's growth trajectory, and to question the logic of more arrivals implying greater benefits' (p. 13).

Although the pandemic has forced nations towards an involuntary downsizing of the tourism industry, this can be regarded as a real-time experiment in downsizing such activity in more normal times, and be used as a catalyst for social change (Fletcher *et al.*, 2020). As Taylor (2020) posted on Facebook, 'the pre-corona existence was not normal other than we normalised greed, inequity, exhaustion, depletion, extraction, disconnection, confusion, rage, hoarding, hate and lack'. While history shows that as soon as the pandemic passes and the growth engine starts, emissions will rise again (Alexander, 2020), now might be a transformational moment that opens up possibilities for rethinking and resetting the tourism industry. However, the question that arises is whether any governments and communities will attempt to avoid a return to a 'high impact, carbon-intensive living, while at the same time ensuring all people feel economically secure in a downshifted economy' (Alexander, 2020). Instead of the urgent need to return to business as usual, when the crisis is over there will be an opportunity to reconsider a transformation to a tourism system more aligned to the degrowth imperative.

As Higgins-Desbiolles (2020) has written about the pandemic: 'after this crisis passes, we might find the old business as usual less compelling. We might learn that not travelling long distances didn't stop us travelling; it just enlivened us to the richness of local travel'. Tourists just need to radically adjust their way of travelling.

In order to recover from the COVID-19 crisis as well as to mitigate climate change and enable human thriving and ecological recovery, the role of domestic tourism is essential. For instance, New Zealand's prime minister has suggested the entire nation move to a four-day week in an attempt to help boost domestic tourism after the crisis (Leasca, 2020). Such an action is in line with the fundamental conceptual and philosophical notions of degrowth which support a sharp reduction in working hours to improve work-life balance and allow more unconstrained time to travel. In addition, the behaviour of tourists in the short and long term should be directed toward avoiding non-essential travel, including business trips, and instead make use of videomeetings, videoconferences, videoconsultations, etc., which will contribute to a reduced use of resources. Also, there is an opportunity for governments to make lasting reforms to the existing model of overconsumption and overproduction and to come up with models that limit 'overtourism'.

To conclude, destinations should become slower by design and not because of economic crisis and/or health and environmental disasters. Policies proposed by degrowth should offer ways for destinations to envision a meaningful transition to a regenerative and more environmentally friendly system and enable tourists to refocus on what really matters by enjoying positive experiences without fear, instead of merely exploiting the natural and sociocultural resources of the host community.

References

Alexander, S. (2020) Is the economic shut down what degrowth advocates have been calling for? Resilience. Available at: https://www.resilience.org/stories/2020-03-24/is-the-economic-shut-down-what-degrowth-advocates-have-been-calling-for/ (accessed 12 June 2020).

Andriotis, K. (2002) Options in tourism development: conscious versus conventional tourism. *Anatolia* 13(1), 73–85. DOI: 10.1080/13032917.2002.9687016.

Andriotis, K. (2005) The Tourism Life Cycle - An Overview of the Cretan Case. *Paper presented at the International Conference on Tourism Development and Planning*, A.T.E.I. of Patra, 11-12 June 2005. Patra, Greece.

Andriotis, K. (2008) Integrated resort development: the case of Cavo Sidero, Crete. *Journal of Sustainable Tourism* 16(4), 428–444. DOI: 10.1080/09669580802154207.

Andriotis, K. (2011) A comparative study of visitors to urban, coastal and rural areas: evidence from the island of Crete. *European Journal of Tourism Research* 4(2), 93–108.

Andriotis, K. (2013) The 'antinomian' lifestyle enclave: an alternative mode of travelling. *Annals of Tourism Research* 40, 40–58.

Andriotis, K. (2018) *Degrowth in Tourism. Conceptual, Theoretical and Philosophical Issues*. CABI, Wallingford, Oxfordshire.

Boonstra, W.J. and Joose, S. (2013) The social dynamics of degrowth. *Environmental Values* 22(2), 171–189. DOI: 10.3197/096327113X13581561725158.

Büchs, M. and Koch, M. (2019) Challenges for the degrowth transition: the debate about well-being. *Futures* 105, 155–165. DOI: 10.1016/j.futures.2018.09.002.

Castaldo, M. (2012) Economy of degrowth and localization: theories and perspectives. BSc thesis. Universit Ca'Foscari Venezia, Venice.

Chassagne, N. and Everingham, P. (2019) Buen Vivir: Degrowing extractivism and growing wellbeing through tourism. *Journal of Sustainable Tourism* 27(12), 1909–1925. DOI: 10.1080/09669582.2019.1660668.

Chen, P.T. and Var, T. (2010) Distribution of tourism economic impacts: a longitudinal study. *International Journal of Tourism Policy* 3(2), 91–112. DOI: 10.1504/IJTP.2010.034206.

Das, D. and Sharma, S.K. (2009) An assessment of the impact of tourism development at Varanasi: perspectives of local tourism businesses. *International Journal of Tourism Policy* 2(3), 167–186. DOI: 10.1504/IJTP.2009.024551.

Eusébio, C., Carneiro, M.J. and Caldeira, A. (2016) A structural equation model of tourism activities, social interaction and the impact of tourism on youth tourists' QOL. *International Journal of Tourism Policy* 6(2), 85–108. DOI: 10.1504/IJTP.2016.077966.

Faccioli, M. (2011) Youth's perception of tourism impact: policy implications for Folgaria (Italy). *International Journal of Tourism Policy* 4(1), 1–35. DOI: 10.1504/IJTP.2011.046710.

Fletcher, R., Murray Mas, I., Blanco-Romero, A. and Blázquez-Salom, M. (2019) Tourism and degrowth: an emerging agenda for research and praxis. *Journal of Sustainable Tourism* 27(12), 1745–1763. DOI: 10.1080/09669582.2019.1679822.

Fletcher, R., Mas, I.M., Blázquez-Salom, M. and Blanco-Romero, A. (2020) Tourism, degrowth, and the COVID-19 crisis. Available at: https://politicalecologynetwork.org/2020/03/24/tourism-degrowth-and-the-COVID-19-crisis/ (accessed 18 July 2020).

Foucault, M. and Miskowiec, J. (1986) Of other spaces. *Diacritics* 16(1), 22–27. DOI: 10.2307/464648.

Gössling, S., Scott, D. and Hall, C.M. (2020) Pandemics, tourism and global change: a rapid assessment of COVID-19. *Journal of Sustainable Tourism* 12(5), 1–20. DOI: 10.1080/09669582.2020.1758708.

Hall, C.M. (2009) Degrowing tourism: décroissance, sustainable consumption and steady-state tourism. *Anatolia* 20(1), 46–61. DOI: 10.1080/13032917.2009.10518894.

Hickel, J. (2015) Forget 'developing' poor countries, it's time to 'de-develop' rich countries. *The Guardian* 23rd September 2015.

Higgins-Desbiolles, F. (2020) The end of global travel as we know it: an opportunity for sustainable tourism. *The Conversation* March 17.

Higgins-Desbiolles, F., Carnicelli, S., Krolikowski, C., Wijesinghe, G. and Boluk, K. (2019) Degrowing tourism: rethinking tourism. *Journal of Sustainable Tourism* 27(12), 1926–1944. DOI: 10.1080/09669582.2019.1601732.

Jackson, T. (2009) *Prosperity without Growth: Economics for a Finite Planet*. Earthscan, London.

Kallis, G., Paulson, S., D'Alisa, G. and Demaria, F. (2020) The case for degrowth in a time of pandemic. Available at: https://www.opendemocracy.net/en/oureconomy/case-degrowth-time-pandemic/ (accessed 4 June 2020).

Lacher, R.G. and Nepal, S.K. (2010) From leakages to linkages: local-level strategies for capturing tourism revenue in northern Thailand. *Tourism Geographies* 12(1), 77–99. DOI: 10.1080/14616680903493654.

Latouche, S. (2007) *La Scommessa della Decrescita*. Feltrinelli, Milan.

Leasca, S. (2020) New Zealand's Prime Minister Just Suggested the Entire Nation Move to a 4-day Workweek. Available at: https://www.travelandleisure.com/travel-news/new-zealands-prime-minister-4-day-work-week?utm_medium=social&utm_content=link&utm_campaign=travelandleisure_travelandleisure&utm_source=facebook.com&utm_term=EB7C008E-9B8B-11EA-9B29-BAA896E8478F (accessed 20 June 2020).

Mas, M., Cadena, Y, I. and Blázquez Salom, M. (2017) *The role of crises in the production, destruction and restructuring of tourist spaces. The case of the Balearic Islands*. University of Alicante, Alicante, Spain.

Mbaiwa, J.E., Motsholapheko, M.R., Kgathi, D.L. and Monnamorwa, S. (2018) Tourism and energy use in lodges and camps in the Okavango Delta, Botswana. *International Journal of Tourism Policy* 8(1), 1–17. DOI: 10.1504/IJTP.2018.090318.

Milano, C. (2017) *Overtourism and Tourismphobia: Global Trends and Local Contexts*. Ostelea School of Tourism and Hospitality, Barcelona, Spain.

Milano, C., Novelli, M. and Cheer, J.M. (2019) Overtourism and degrowth: a social movements perspective. *Journal of Sustainable Tourism* 27(12), 1857–1875. DOI: 10.1080/09669582.2019.1650054.

Monterrubio, J.C. and Andriotis, K. (2014) Social representations and community attitudes towards spring breakers. *Tourism Geographies* 16(2), 288–302. DOI: 10.1080/14616688.2014.889208.

Scheyvens, R. (1999) Ecotourism and the empowerment of local communities. *Tourism Management* 20(2), 245–249. DOI: 10.1016/S0261-5177(98)00069-7.

Taylor, R.S. (2020) We will not go back to normal. Available at: https://www.sonyareneetaylor.com (accessed 20 April 2020).

Valdivielso, J. and Moranta, J. (2019) The social construction of the tourism degrowth discourse in the Balearic Islands. *Journal of Sustainable Tourism* 27(12), 1876–1892. DOI: 10.1080/09669582.2019.1660670.

Index

Note: Page numbers in **bold** type refer to **figures**
Page numbers in *italic* type refer to *tables*